Design for
INHERENT SECURITY

Guidance for non-residential buildings

Barry Poyner and William Fawcett

CONSTRUCTION INDUSTRY RESEARCH AND INFORMATION ASSOCIATION
CIRIA Special Publication 115

Published by Construction Industry Research and Information Association,
6 Storey's Gate, Westminster, London SW1P 3AU

© CIRIA 1995

ISBN 0-86017-416-6

Book design and page layouts by Cambridge Architectural Research Ltd

SUMMARY

The inherent security of a building and its occupants depends on site layout, the planning of the building, and its detailed design. These are more or less permanent features and defects in inherent security cannot be 'fixed' by add-on devices. Inherent security may cost little or nothing, so long as it is allowed for at the earliest stages of a project. This guide explains the principles of design for inherent security, as applied to non-residential buildings.

Where possible the advice is based on research findings, but because research on this topic remains sparse the findings have been augmented by what seem to be reasonably reliable current practices.

As well as helping architects, building owners and users in the design of new or refurbishment projects, it is intended that this guide will increase awareness of the all-pervasive relevance of security in design, and stimulate further research into inherent security.

B POYNER and W FAWCETT
Design for Inherent Security: guidance for non-residential buildings
Construction Industry Research and Information Association
Special Publication 115, 1995

Keywords:
Security, crime, crime prevention, architectural design, design briefing

Reader interest:
Architects, surveyors, facilities managers, building owners and developers, security managers, crime prevention officers

CLASSIFICATION	
AVAILABILITY	Unrestricted
CONTENT	Subject area review
STATUS	Committee guided – Final
USER	Architects, other designers, building clients, security/crime prevention personnel

FOREWORD AND ACKNOWLEDGEMENTS

This guide was prepared by Barry Poyner of Poyner Research Consultancy and Dr William Fawcett, Dr Andrew Coburn and Dr Stephen Platt of Cambridge Architectural Research Ltd, under contract to CIRIA.

Following CIRIA's usual practice the research was guided by a Steering Group, which comprised:

P Veater (Thorn UK Ltd, Marlow) – *Chairman*

Dr E C Bell (Loss Prevention Council, London)

M Buckley (The Association of Burglary Insurance Surveyors Ltd)
  – *for first part of project*

Professor R Burgess (Burgess Jepson & Co Ltd, Merseyside)
  – *representing the Chartered Institute of Building*

Dr A J Burns-Howell (Dixons Group Plc, London)

R Carter (Modern Security Systems Ltd, Salford)

J Finney (British Security Industry Association, Worcester)

Sergeant P Hardy (Sussex Police)

Chief Inspector B Hewitt (Home Office Crime Prevention Centre, Stafford)

D Hincks (TBV Consult, Croydon)

N Newton (Guildford College of Further and Higher Education)
  – *representing the Chartered Institute of Building Services Engineers*

T Pascoe (Building Research Establishment)

A meeting of the project Steering Group at CIRIA's office in London

R Reed (formerly of Legal & General Insurance, London)
– *representing the Association of British Insurers*

G Squire (Henry Squire & Sons, Willenhall)
– *representing the British Hardware & Housewares Manufacturers' Association*

D Williams (Ellis Williams Partnership, Manchester and London)

P Woodhead (Department of the Environment)

CIRIA is most grateful for their contribution to this project.

CIRIA's Research Manager for the project was Mrs Ann Alderson.

The project was funded by:

British Hardware & Housewares Manufacturers' Association
Building Centre Trust
Department of the Environment
Dixons Group Plc
Home Office
Interbuild Fund
Loss Prevention Council
Modern Security Systems Ltd
Thorn UK Ltd

In the course of the project two research workshops were held. The participants in the first workshop, held in June 1993, were:

J Bourke (Cambridge)
R Brimblecombe (Cowper Griffith Brimblecombe, Cambridge)
A Chorley (Anthony Chorley : Architects, Warley, W Midlands)
A Cooper (Rickaby Thompson Associates, Cambridge)
J Crosby (Bolton & Crosby Architects, Harrogate)
K Davis (Bader Miller Davis Partnership, London)
I C Farquhar (Abacus Architects Ltd, Chelmsford)
I McGregor (Elder Lester Associates, Yarm, Cleveland)
C B Nicholson (Clive Nicholson Associates Ltd, Cambridge)
D Pattenden (Suffolk County Council, Ipswich)
A Procuta (Annand & Mustoe Associates, Cambridge)
R Robinson (The Charter Partnership, Bedford)
R Saunders (Saunders King Moran, Ashton-under-Lyne, Lancs)
B G Woodcock (Woodcock Associates, Bridlington, E Yorkshire)

The participants in the second workshop, held in February 1994, were:

C Atkins (RDA Architects, Godalming, Surrey)
P Bradley (Damond Lock Grabowski & Partners, London)
D Brooke (Bedwell Brooke Partnership, St Albans, Herts)
J M G Fox (Rolfe Judd, London)
J Devas (Scott Brownrigg & Turner, London)
R Heath (Douglas Feast Partnership, London)
M Irwin (John Living & Morrison Rose Architects, London)
J Jenkins (Thames Valley Police)
J Lapthorne (GMW Partnership, London)
J Shorten (Surrey Police)
B Sopp (Oxford Architects Partnership)
C Stagg (Metropolitan Police)
A Starkey (Cecil Denny Highton, London)
D Stephens (Stillman Eastwick Field Partnership, London)
R S Thelwell (Hunter & Partners Architects Ltd, London)
P M Ward (JKL Architects & Planners, Basingstoke, Hants)
J White (Hertfordshire Police)
P Wilgoss (Metropolitan Police)

CIRIA, Barry Poyner Research Consultancy and Cambridge Architectural Research Ltd gratefully acknowledge their contribution to the development of this guide.

The following photographs are reproduced with permission:

page 33 – Laminated Glass Information Centre
page 59 – Euroquipment Ltd
page 66 – Leabank Office Equipment Ltd
page 67 – Dexion Ltd
page 70 – Knoll International Ltd
page 71 – Zon International Ltd
page 77 – Otto Wöhr Gmbh Parksysteme
page 79 – Henry Squire & Sons
page 91 – Gunnebo Protection Ltd
page 95 – Bewator (UK) Ltd
pages 106 and 107 – Heras Fencing Systems (UK) Ltd
page 137 – Bonar & Flotex Ltd

CONTENTS

Part 3 · Security issues in selected building types

Part 4 · Notes on security measures

Part 5 · Further information

Design for
INHERENT SECURITY

INTRODUCTION

This publication provides guidance to architects and building owners or users on security in the design or refurbishment of non-residential buildings. Its main purpose is to raise awareness of security as one of the fundamental performance requirements of buildings. The security issues presented in this guide affect almost every aspect of building layout and design and need to be considered early in the design process.

Scope and objectives

Security problems affecting the design of non-residential buildings have received less attention from research than those affecting housing. The purpose of this guide is to draw together the findings of research into security in non-residential buildings and to present guidance for designers and their clients.

There are good reasons for believing that security in non-residential buildings presents a somewhat different and certainly a more diverse set of problems for the designer. The most obvious difference is that non-residential buildings, while well supervised during working hours, are often unoccupied at night, whereas the reverse is generally true for housing. Another difference is that non-residential buildings are often used by large numbers of people and many are accessible to the general public.

The variety of security problems in non-residential buildings is enormous. For this reason the guide does not attempt to cover every aspect. It concentrates on the main security issues and the principal methods of tackling them through design.

Research basis

The intention throughout the preparation of this guide has been to rely on the results of research. It is believed that this is the only solid basis for making design recommendations. However, it has become increasingly clear that research studies have hardly scratched the surface and that guidance which was restricted to matters covered by research would be too patchy to be useful. So it was decided to include some elements from established good practice, where it appears to be consistent with research. The authors have tried to indicate where the recommendations are based on research.

It would be desirable if the publication of this guide could, as a secondary objective, draw attention to the need for more research into the relationship between security and design.

Security issues, not security measures

One important innovation in this guidance is the emphasis on *security issues*, not security measures. Security issues, as used here, have three important features:

● they are generic and common to many different building types

● they integrate crime problems and design principles

● they have a concise, modular form.

Some of the issues will be familiar to everyone, for example 'Common burglary' (§1). But others, such as 'Customer only access' (§15), will probably be unfamiliar, even to those who have read the existing security literature. This is because the concept of security in this guide is broader than the traditional view of designing for security.

No attempt has been made to rank security issues in order of importance. The main reason is that the circumstances of individual cases will determine which issues are applicable and their relative importance. A second reason is that designers should be aware of all security issues, not just the 'important' ones.

The guidance draws primarily on research which relates the incidence of crime to design features in the environment. It does not focus on the related but different topic of fear of crime and how this is affected by design.

How the guidance is organised

Security is discussed at a strategic level. Detailed guidance on particular security measures can already be found in other publications which are referenced at appropriate points.

There are five parts:

Part 1, *An approach to building security*, sets out the general concept of inherent security in building design. It emphasises that security is one of the fundamental aspects of building performance and explains how the idea of inherent security can be incorporated in the design process.

Part 2, *Security issues in building design*, is the heart of the guide. The *master list of security issues* provides a summary or checklist of the 21 security issues. It is followed by a fuller presentation of each issue using broadly the same structure, beginning with a description of the crime or security threat, followed by an indication of the places where it is likely to occur; then the approaches to prevention that could overcome or reduce the problem, followed by a number of design principles; and finally there are references to research or other sources of further information. Note that the design principles are neither detailed nor exhaustive – they are intended as the starting point for design.

Part 3, *Security issues in selected building types*, relates the security issues to a number of common non-residential building types. For each building type there is a checklist of security issues, supplemented where appropriate by summaries of research findings.

Part 4 contains *Notes on security measures*. Without duplicating the detailed coverage that is already available in other sources, it introduces and provides a commentary on some of the security measures that are commonly recommended for security design.

Finally, Part 5 references sources of *Further information*.

How to use the guide

This guide is intended for browsing and for reference – few readers, if any, will read it from cover to cover.

When used as a reference source in the context of a particular design project, an outline design procedure is described in Part 1 (see page 15).

The guide should not be restricted to a reference for particular projects. When evolving basic design concepts, designers use ideas that they have assimilated already. If the ideas of inherent security are to have an impact at concept stage, it is important they become established in designers' minds. So it is hoped that designers will find time to browse through the whole document, and in particular read Part 1, *An approach to building security*.

■

PART 1 AN APPROACH TO BUILDING SECURITY

AN APPROACH TO BUILDING SECURITY

The design of buildings can provide an important means of controlling crime and nuisance behaviour. Traditionally, security in the design of buildings has been seen as the problem of preventing burglary. However research, much of which will be referred to later in this guide, has shown that the implications are much wider. Many types of crime and nuisance behaviour can be controlled through building design.

Preventing crime

Don't attempt to reform man...

My philosophy and strategy confine the design initiative to reforming only the environment in contradistinction to the almost universal attempts of humans to reform and restrain other humans by political actions, laws, and codes.

R Buckminster Fuller *Utopia or Oblivion* (1969)

In recent years much attention has been given to crime prevention through the design of buildings and the environment as part of a wider policy of preventing crime. There has been a good deal of design guidance published on building security, particularly for housing. The police have become involved in advising architects on 'designing out crime' (Clarke and Mayhew, 1980). The idea has also spread to the planning process where authorities are encouraged to consider the implications for crime of new developments (Department of the Environment, 1994).

However, not everyone agrees that environmental design has a significant role in preventing crime. Many criminologists and social scientists point out that crime has its roots in social, cultural and economic problems. They insist that crime prevention must deal with root causes through improvements in educational, social and economic policies. The social approach to crime prevention also includes a wide range of local projects to encourage young people to find acceptable outlets for their energies.

The prevention of crime relies on the combination of social prevention, the use of the police and criminal justice system, and environmental design. In this context environmental design includes all aspects of the design and management of the physical environment.

Others put their trust in the police and the criminal justice system. They believe that the solution to the problem of crime lies in increasing the powers of the police in detecting criminals and bringing them to justice. For some this implies harsher justice and stronger punishments as a means of deterring the criminals.

Arguments have raged to and fro over these different positions. Those who advocate social methods of prevention, rather than changes to the physical environment, often cite the theory of displacement. This theory says that changing part of the environment cannot prevent crime because it is simply displaced to other, less well defended, targets.

Similarly, arguments against strengthening the police and criminal justice system point out the ineffectiveness of increased prison sentences or other harsh treatment regimes in reducing re-offending by criminals.

The solution to the problem of preventing crime probably lies somewhere in the middle. Few will deny that the kind of start children have in life and the way they are educated has an effect on how they behave. Similarly, although there are usually alternative targets for a determined criminal, there is ample evidence to demonstrate that design changes can reduce the opportunity for crime and bring about a general reduction in criminal activity (examples are given in Clarke, 1992). Furthermore, these environmental design measures will only act as a deterrent if criminals face a reasonable risk of being caught and punished. For this to be the case there must be effective police and criminal justice systems.

A new approach

No amount of intruder detection equipment or other 'hardware' can compensate for a building or site which is fundamentally insecure.

N Cumming *Security* (1992)

The conventional approach to improving building security has been to fit better locks on doors and windows and to add security devices such as bars and shutters. Other familiar types of security equipment include alarm systems, closed-circuit television and security lighting.

All of these measures have their place in providing good security, but the idea that improved security is purely a matter of adding more and more security equipment to the designed environment is being questioned. There are a number of reasons for this.

Aesthetic concerns

First there are aesthetic concerns about an environment dominated by security paraphernalia. A topical example is in the objection by many planning authorities to the extensive use of security shutters in shopping streets after trading hours (Shop Front Security Campaign, 1994). Such security measures are seen as damaging to the quality of urban life.

Many are also unhappy about the increasing use of measures such as controlled entry systems, alarms and security cameras which reduce freedom of movement and privacy. Some argue that these overt security measures tend to raise the level of fear of crime and anxiety over personal safety.

Effectiveness of security measures

There are also doubts about the effectiveness of many security measures. It has been argued, for example, that improved street lighting will reduce crime rates, but researchers from the Home Office have found little firm evidence for this (see 'Lighting and security' in Part 4, page 158).

Another problem has been that stronger security measures lead to more forceful methods being used by criminals. The window lock was introduced to prevent a common means of entry through windows by breaking the glass and reaching in to open the catch. Now that window locks are more common, the method of entry is more usually by forcing the window, using a simple hand tool.

A more dramatic example of the escalating use of force is the ramraid. As retail premises have been increasingly protected by alarm systems, reinforced doors and shutters, some criminals have raised the stakes. Instead of using stealthy methods of entry, they now use stolen vehicles to smash a way into the building, grab what they can and escape in another stolen vehicle before the police can arrive.

In the Front and Sides, as he cannot always be sure of having honest neighbours, he must make his Walls stronger against the Assaults of both Men and Weather.

Alberti *The Ten Books of Architecture* (1472) – the C16 rustication is by Sangallo

The cost of security technology

In addition to concerns about the effectiveness of security measures, there is the problem of cost. The more sophisticated the equipment, the greater the cost of installation and the greater the cost of support and maintenance. Modern security systems, such as intruder detection alarms and surveillance cameras, are only fully effective when they have human back-up. This may require a continuously manned monitoring station to ensure a fast response by the police. Even this may be insufficient, requiring the presence of full-time security personnel on site.

Another aspect of cost is that where security installations are damaged in a criminal attack, the cost of repair may exceed the losses from theft.

Security through design

Since the early 1970s there has been a growing interest in the idea that crime can be controlled more effectively through the general design of a building than through added security technology. Oscar Newman's well-known 'defensible space' proposals, although criticised by some researchers, drew attention to the idea that some housing designs were more defensible than others (Newman, 1973).

Newman pointed out that defensible space could be achieved through design by deliberately defining territory in and around housing blocks which could be watched and controlled by the residents themselves. His work was influenced by the earlier writings of Jane Jacobs, who had drawn attention to the importance of a diversity of businesses and residences facing onto the street in achieving safe urban environments (Jacobs, 1961).

The idea of crime prevention through environmental design (CPTED), a phrase first coined by C Ray Jeffrey (1971), grew to some prominence in the United States during the 1970s (Crowe, 1991). Although there are a number of different formulations of this approach to crime prevention, essentially it aims to control crime by managing patterns of access and movement, and by

Public buildings were traditionally massive, intimidating and inaccessible, for practical and symbolic reasons – but architects of the Modern Movement rejected this tradition: compare museums of the 1830s and 1960s (Fitzwilliam Museum, Cambridge, by Basevi, top; and National Gallery, Berlin, by Mies van der Rohe).

A city street equipped to handle strangers, and to make a safety asset out of the presence of strangers ... must have three main qualities:

First, there must be a clear demarcation between what is public space and what is private space...

Second, there must be eyes upon the street, eyes belonging to those we might call the natural proprietors of the street...

And third, the sidewalk must have users on it fairly continuously...

Jane Jacobs *The Death and Life of Great American Cities* (1961)

providing surveillance in and around buildings from the occupants and other users. A corollary is that in planning buildings it might be possible to arrange activities in such a way that they provide mutual surveillance or protection.

Since these ideas were developed in the 1970s there has been a continuing research interest in the concept of controlling crime through design. Most of this effort has been focused on housing (for example, Poyner and Webb, 1991). The purpose of the present guide is to bring together research and to identify guidance about design as a means of controlling crime and nuisance behaviour in non-residential environments.

The concept of inherent security

Work on this guide has given rise to the concept of *inherent security* in building design. Inherent security is not achieved by security equipment and alarm systems, but through physical layout and the location of doors and windows, the control of movement and access, and the exploitation of natural or informal human surveillance. These are exactly the elements that the architect already manipulates in every design project: inherent security does not, therefore, introduce new elements into design. By manipulating the familiar elements of architectural design with a new awareness of their impact on security, there is a potential for achieving significant gains in crime prevention and user satisfaction.

Inherent security can provide a good basic standard of security for all buildings, but by itself it will not give protection for very high risk activities, such as banks or post offices or indeed any premises where high value goods or materials are processed or stored. However, it is suggested that buildings designed with good inherent security provide a sound basis for upgrading through the addition of appropriate security devices and management procedures.

An economic solution to security

A major factor supporting this approach to security is its cost implications. Inherently secure designs should cost less than conventional designs plus additional security equipment and support systems. Furthermore, it is reasonable to believe that owners and tenants of buildings that are well designed in terms of inherent security are less likely to have to spend considerable sums of money installing security systems to make up for poor design characteristics. Studies indicate that a major element of the cost of crime to businesses is in the cost of preventive measures they have to adopt. For example, it has been estimated that the retail industry spent £370 million on crime prevention measures in 1992/93, which was 19% of the total cost of retail crime (Burrows and Speed, 1994). Well designed buildings should reduce this burden of expenditure and reduce the extent to which buildings would need to be retrofitted with security equipment.

Scope of inherent security

In compiling this guide the aim has been to review the research literature to identify what aspects of design appear to influence crime and related nuisance behaviour. Although it was clear that some crimes, such as fraud, were unlikely to be influenced by building design, the more the review progressed, the more it became clear that there were many security problems which could be improved through design. Indeed, almost everything in design – from the arrangement of buildings on a site, the positioning of walls, windows, doors, fences, screens, down to the use of lighting, bollards, landscaping and so on – can have security implications.

Design for inherent security can be broadly defined as the use of the site layout, physical form and fabric of a building to protect the occupants and their possessions from harm which may be caused by fellow human predators. This includes all forms of theft, violence and damage. Put in this way, security is shown to be a basic human need in building design, alongside shelter from wind, rain, heat and cold and requirements such as privacy, hygiene, light and good acoustics.

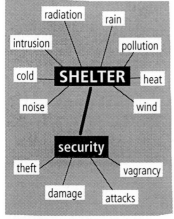

Security from all forms of theft, violence and damage is a basic human need in building design. It stands alongside shelter from wind, rain, heat and cold and requirements for privacy, hygiene, etc.

Inherent security in the design process

In the conventional approach to design for security, which typically relies on the specification of doors, architectural ironmongery, alarms and surveillance equipment, designers leave security to be dealt with after the basic *design concept is established.*

However, the concept of inherent security goes to the very heart of the design concept and must be considered at the earliest stage of design. Inherent security is largely determined by the basic design concept and cannot be added on later.

What follows is an outline procedure for analysing the security requirements in a design project. The approach applies equally to the re-modelling of an existing property as to the design of a new development.

Identifying security risks

The approach adopted in this guide is to classify the various risks which building designers might address. They are presented as *security issues* – a term used to describe both the security risk or crime problem together with the design principles which enable it to be controlled.

The task at the very earliest stages of briefing is to make an assessment of the security risks for the project. To help designers and their clients do this, a *master list of security issues* is presented at the beginning of Part 2 (page 22). It is intended as a checklist for identifying which issues might be relevant. The supporting presentation for each issue suggests approaches to prevention and design principles that could be applied in any given project. Part 3 gives checklists of security issues for some common building types, and summarises relevant research studies where they are available.

In assessing the risks that need to be considered in design, there are a number of resources which should be drawn on. If the building is to be owner-occupied or if the future tenants are known, it may be that these organisations have their own security managers with experience and data on the risks associated with their activities. Clients' security

LET US STATE THE PROBLEM ... a shelter against heat, cold, rain, thieves and the inquisitive. A receptacle for light and air. A certain number of cells appropriated to cooking, work and personal life.

Le Corbusier *Towards a New Architecture* (1927) – the building is Le Corbusier's Zurich Pavilion of 1963–67

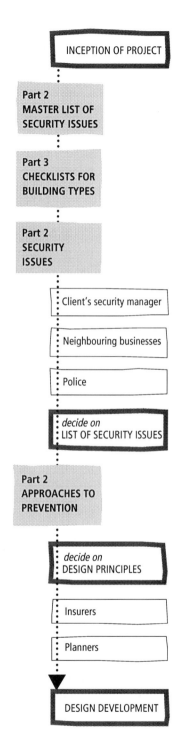

INCEPTION OF PROJECT

Part 2
MASTER LIST OF
SECURITY ISSUES

Part 3
CHECKLISTS FOR
BUILDING TYPES

Part 2
SECURITY
ISSUES

Client's security manager

Neighbouring businesses

Police

decide on
LIST OF SECURITY ISSUES

Part 2
APPROACHES TO
PREVENTION

decide on
DESIGN PRINCIPLES

Insurers

Planners

DESIGN DEVELOPMENT

managers or advisors would be the first source of specialist knowledge to call on.

If the site is in an area where other similar businesses or developments operate, it may be wise to find out what experiences they have of crime-related problems. This may be particularly relevant for aspects of public order and nuisance, such as rowdiness, vagrants, and vandalism.

The next resource is the local police. All police forces in England and Wales have specially trained officers to provide advice to architects and developers. They are known as Architectural Liaison Officers or Crime Prevention Design Advisors. Part of their task is to provide crime intelligence about the risks of particular sites based on existing crime data. They should also be aware of local problems which would not necessarily appear in crime statistics and which are not covered in Part 2, such as the management of crowds at nearby sports grounds or seasonal problems like 'new age' travellers. Their advice is offered without charge to designers and building owners.

Two further sources of advice are insurers and planners. Early consultation with insurers may reduce the need to add expensive security measures at a late stage of design or after the building is occupied. Planning authorities are beginning to develop policies with security implications, encouraged by the recent Government circular, *Planning out Crime* (Department of the Environment, 1994).

The design response

The purpose of the guidance given in Part 2 on security issues is to show how design can be used to minimise or reduce the security risks that are identified in the brief. Each section describes a security issue in some detail and discusses the approaches which can be adopted for the prevention of the crime or nuisance concerned. Physical design is not the only means of controlling crime and in practice most security problems will be managed by a combination of operational policies and design.

Each security issue is written in such a way that it leaves it open to designers and their clients to decide the best approach for their particular circumstances. Only when this decision is made will it be clear which of the design principles should be adopted.

The current state of knowledge in this field makes it impossible to determine precisely the extent to which the design principles will reduce crime risks. At present all that seems reasonable to claim is that where the principles are successfully incorporated into design, the design will provide a far more secure environment.

It is common in security management to use the terms 'risk management' and 'risk assessment' in a rather precise way. The main reason is to support the financial appraisal of security measures. This is less relevant to inherent security because for the most part the design principles do not incur specific costs. For example, arranging a building plan so that office windows overlook a car park need not involve extra cost, whereas the installation of a CCTV surveillance system or the employment of a security guard does have a direct cost.

However, risk assessment becomes more relevant where security risks are judged to be beyond the scope of inherent security. This will arise for high risk activities and it is clear that in-house or specialist security advice will be necessary.

The purpose of this guide, therefore, is not to provide detailed guidance on every aspect of security in buildings. It is primarily intended to help designers and their clients improve the inherent security of their buildings. It is written in the belief that non-residential buildings will be significantly less vulnerable to crime and other nuisance behaviour if the design principles set out in Part 2 are followed. Furthermore, where buildings are designed with inherent security in mind, the cost of security measures and the need for retrofit to deal with security problems will be greatly diminished.

No method of design ... can assure that there is no risk of failure under any circumstances that can arise. All that any design method assures is that the chance of failure is sufficiently low.

Building Research Station
Principles of Modern Building (1959) – discussing structural design

Where to find out more

Burrows, J and Speed, M (1994)
Retail Crime Costs 1992/93

Clarke, R V and Mayhew, P (1980)
Designing out Crime

Clarke, R V (ed) (1992)
Situational Crime Prevention: successful case studies

Crowe, T D (1991)
Crime Prevention through Environmental Design

Department of the Environment (1994)
Planning out Crime

Jacobs, Jane (1961)
The Death and Life of Great American Cities

Jeffrey, C Ray (1971)
Crime Prevention through Environmental Design

Newman, Oscar (1973)
Defensible Space: people and design in the violent city

Poyner, B and Webb, B (1991)
Crime Free Housing

Shop Front Security Campaign (1994)
Shop Front Security Report

■

If I play chess against a master he will always win, precisely because he can predict my behavior while I cannot do the reverse... Note that the chessmaster does not control my behavior by any psychological device. He simply makes legal moves, which have the effect of changing my environment and the possibilities it offers. Indeed, this is almost always how behavior is contolled. Changing the world is a very powerful way of changing behavior; changing the individual while leaving the world alone is a dubious proposition.

U Neisser *Cognition and Reality* (1976)

PART 2

SECURITY ISSUES IN BUILDING DESIGN

MASTER LIST OF SECURITY ISSUES

The master list summarises 21 common security issues which can arise in the design of non-residential buildings. The issues are covered in more detail in the following sections in Part 2. The master list can also be used as a checklist during the briefing and design of any non-residential building. Each security issue has a code number which is used throughout the guide.

The master list is intended to cover most of the common security issues in the design of non-residential buildings, but it is not exhaustive. It does not cover high security risks, such as in the retailing of high value commodities or the storage of art treasures. Nor does it deal with the more specialist forms of security, such as the detention of prisoners, or with terrorism or matters of national security.

Some building uses create special security issues. Examples include the risk of the abduction of babies from hospital maternity units or the protection of toxic biological or chemical substances at research laboratories. Such issues can only be identified through careful discussion between designers and their clients at an early stage in the development of the design brief.

It should be stressed that not all crime can be controlled through design. For example, fraudulent business activities are unlikely to be affected by design. Other areas of criminal activity which are of concern but seem unlikely to have implications on building design include computer crime and sexual harassment of staff.

FORCED ENTRY	**1 Common burglary** The most common form of burglary is breaking into a building without being seen or heard for the purpose of theft. Common burglaries usually take place when buildings are unoccupied. **2 Smash and grab raids** Smash and grab raids are attacks on display windows or showcases. The raiders are not concerned about creating a noise or setting off an alarm. The essence of these attacks is surprise and a quick getaway. **3 Ramraids** Ramraiding is an extreme version of smash and grab raids using a stolen vehicle to smash a way into the building. Despite its violence, it does seem to be a type of crime that can be readily controlled through design.
VIOLENCE	**4 Cash robbery** Wherever business transactions involve cash, there is a risk of staff being attacked or threatened with violence in attempts at robbery. **5 Violence to staff** The problem of protecting employees from violence other than robbery. It is predominantly a problem of verbal abuse, but physical violence does occur.
THEFT	**6 Shoplifting** Unlike other security issues, shoplifting is by definition specific to one type of environment – shops and retail stores. It involves the theft of displayed goods by customers or individuals posing as customers. **7 Employee pilferage** Unfortunately, not all employees can be trusted with the goods and supplies which they must handle during their work. Although most control systems have little to do with building design, there are some design principles which may help. **8 Customer belongings** Wherever members of the public are receiving a service and have to be temporarily separated from their belongings, there is a need to protect their property from theft or damage. **9 Personal belongings at work** The problem of providing staff with safe places to keep clothing, shopping, other personal belongings, and especially valuables such as wallets and purses, during working hours. **10 Other thefts from premises** Wherever buildings are accessible to the public there is a risk of portable materials, supplies, furnishings, etc, being stolen.

VEHICLE SECURITY	**11 Parked cars**

11 Parked cars

Whenever cars are left unattended in car parks there is a risk that they may be stolen, broken into, stripped of items such as wheels, or simply damaged in some malicious way.

12 Bicycle parking

This is the problem of providing safe places to leave bicycles. Although environmental policies (to reduce pollution and transport congestion) favour bicycles, fear of theft discourages their use.

13 Loading and unloading

The areas around loading and unloading points should be designed to reduce the risk of theft, pilferage or even attacks.

14 Fly-parking

If a private car park or even a landscaped area is located where there is pressure on parking space, there is a need to exclude unwanted cars.

ACCESS

15 Customer only access

There are many situations where businesses or public facilities need to welcome customers and visitors but discourage or prevent unacceptable use and activity.

16 Staff only access

A common problem is managing access to a building or parts of a building in order to provide access for staff while excluding the general public, customers or unwanted intruders.

17 Fly-tipping

Areas of open land, whether landscaped or unused, can be illicitly used for tipping waste, resulting in loss of amenity, nuisance and sometimes pollution.

18 Loitering groups of youths

Groups of youths loitering in public areas can be a source of fear to some members of the public and are often considered to be undesirable for business reasons.

DAMAGE TO BUILDINGS

19 Wilful damage

Damage to buildings can be caused by accident or wear and tear, but sometimes it is the result of a wilful act which might be prevented.

20 Graffiti

Some people find graffiti 'art' beautiful but it is a nuisance to most building owners. Worse still, it can signal decline and neglect and thereby invite other crime or nuisance behaviour.

21 Arson

Much fire-setting by children and youths is similar to vandalism and the risk can be controlled to some extent through design. ■

1   COMMON BURGLARY

The most common form of burglary is breaking into a building without being seen or heard for the purpose of theft. Common burglaries usually take place when buildings are unoccupied.

The problem

Burglary is the crime most people associate with building security, but its legal meaning in England and Wales is very wide. It can include any entry into a building, forced or otherwise, if there is an intention to commit a crime (theft, damage, violent assault or rape). Here the focus is on burglary for theft.

Most burglars prefer to avoid any confrontation and so attempt to break into premises unobserved and without raising any alarm. Only when this approach to burglary is made very difficult and where there are highly attractive targets will thieves turn to more forceful or destructive methods of entry, such as smash and grab raids (§2) and ramraiding (§3), or even adopt the more direct confrontational approach of robbery (§4).

Because an essential characteristic of common burglary is the avoidance of being seen acting suspiciously, the burglar carries only light equipment such as hand tools. Similarly, to avoid being heard, only limited amounts of force are used and noisy methods of entry, such as smashing large areas of glass, are avoided.

The burglar may gain access through any weak point in the building envelope which can provide an opening large enough for a person (sometimes a child) to climb through. The usual points of access are doors or opening windows which are left insecure or forced open. Other weak points

CIRIA SPECIAL PUBLICATION 115 · DESIGN FOR INHERENT SECURITY · 1995

can include fixed windows, rooflights, insubstantial types of cladding and service ducts.

In some more sophisticated burglaries alarm systems are overcome. Most commonly, the alarm is isolated by cutting telephone wires and disabled by filling the outside bell with foam and smashing the strobe light. In some cases the alarm is circumvented by use of inside information or careful observation. Other techniques have been used, such as repeatedly setting off alarms until the police and owners either ignore or disconnect them.

Where the risk occurs

Burglary is a relatively widespread crime. The number of burglaries in non-residential buildings recorded by the police in England and Wales is now more than 650,000 per year. Most of these will be common burglary.

All buildings are potentially at risk from common burglary, but three important factors seem to increase the risk.

The first is the attractiveness of the building's contents. The aim of common burglary is theft and any building which is likely to contain cash or goods which can be readily re-sold for cash is an obvious target. This explains the attractiveness of retail premises and some distribution warehouses. Other building types, such as offices, are attractive for any cash or valuable equipment like desktop computers which can be re-sold. Sometimes document-ation has value, such as blank cheque books and other official forms. Schools are targeted for their sports and computing equipment.

Premises with high value contents like banks, building societies and retail and warehousing premises for high value commodities present higher risks. These may not be adequately protected by the design principles in this section.

Office windows facing directly onto a suburban high street appear to be vulnerable – but it is more likely that burglars would try to gain entry from the car park at the back.

Two other risk factors have been identified by research on burglary (see, for example, Bennett and Wright, 1984). They are a building's occupancy and surveillance. Although the research is mainly derived from residential burglary, it is clear that burglars are more likely to attempt to enter buildings that are unoccupied. This is one reason why non-residential buildings are most at risk outside normal working hours, such as at night and over the weekend.

The third factor is surveillance. Burglary attempts are much less likely in buildings or parts of buildings that are subject to surveillance, which is largely determined by the extent to which the building is visible from the street or overlooked from other buildings. This explains why some more isolated buildings such as schools experience high levels of burglary.

Approaches to prevention

Although proposals for the prevention of burglary usually focus on reinforcing the building envelope, risk management approaches can also be taken. One is to make the potential targets of theft less attractive. Methods include the marking or labelling of goods and equipment which render them much less easy to re-sell. Making objects less portable reduces the scope for burglary, so equipment that is heavy or fixed is less at risk, as are goods in bulk storage systems, such as palletisation.

Another approach might be for businesses with high risk items on their premises to operate with a low profile. This approach is sometimes adopted in transport with the use of unmarked vehicles. Obviously this strategy has only a limited application and makes little sense for businesses which thrive on high-profile advertising, such a retailers, but it might be effective for research or repair and maintenance facilities.

The most direct approach to preventing burglary is usually seen as the upgrading of physical security measures, with locks, bolts and bars, etc. Whilst physical protection is important, it can only be truly effective in the context of some means of human intervention or guardianship. However well protected a building is physically, unless there is some prospect of human intervention burglars can

Take Care, too, that the Windows and Doors do not lie handily for Thieves.

Alberti *The Ten Books of Architecture* (1472)

CIRIA SPECIAL PUBLICATION 115 · DESIGN FOR INHERENT SECURITY · 1995

set to work on breaking in, taking as much time as they like and using noisy drills or cutting equipment.

It is because all physical security measures can be overcome that occupancy and surveillance – which can be thought of together as 'guardianship' – are so relevant to design for the prevention of burglary.

One approach to maximising guardianship is to provide a 24-hour staff presence. This occurs as a matter of course in a few situations, such as a hospital emergency department or an all-night motorway service station. Alternatively, full-time security staff are often employed in buildings like industrial premises. Where security staff are employed there may be a case for CCTV surveillance. In these cases, physical measures may be less important.

A much greater problem arises in buildings which are left unoccupied. Research findings, mainly from housing (Poyner and Webb, 1991), indicate that the most effective approach to burglary prevention is a combination of moderate physical security measures with some natural surveillance from the street and surrounding properties. This preferred situation is achieved largely by careful attention to building layout and site planning.

Design principles

Where a building does not have continuous occupancy or guardianship, but does have the potential for natural or informal surveillance, the following design principle applies.

Natural surveillance

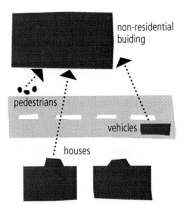

non-residential buiding

pedestrians

vehicles

houses

Buildings should face busy streets or be overlooked by surrounding properties that are likely to be occupied at times when the building is unoccupied.

For effective natural surveillance the building should not be screened by heavy landscaping or high walls or fences – barriers above about 1m high should be transparent, for example using wire mesh or railings.

Good natural surveillance is provided by activities that continue throughout the period when a building is unoccupied. Examples include well-used thoroughfares and pedestrian areas; mixed uses such as restaurants, sports and entertainment facilities alongside premises only open during conventional office hours; businesses on industrial estates with continuous working patterns; and any 24-hour facilities such as petrol stations, emergency services, transport facilities, taxi businesses, hotels with night porters, etc.

Lighting assists natural surveillance during the hours of darkness. Areas covered by street lighting do not usually need additional lighting. Unlit areas where there is natural sur veillance benefit from additional lighting. Well distributed low intensity lighting is more effective than patches of high intensity lighting. Some retail and commercial premises leave internal ground level lighting on at night. (See also 'Lighting and security' in Part 4, page 158.)

Doors, windows or other potential access points which have good informal surveillance are not expected to suffer sustained, heavy attacks from burglars, so they should not require exceptional physical protection. BS8220 and similar guides set out basic methods of security protection. (See also 'Windows and burglary' and 'Security of doors' in Part 4, pages 149 and 153 respectively.)

Alarm systems may be a useful addition to the protection offered by informal surveillance, particularly where surveillance is uncertain and the building contains higher risk contents. Insurers often regard alarm systems as a basic requirement in commercial buildings.

Where a building, or part of a building, is not subject to natural surveillance it must be designed to withstand attempts at illegal entry. There are three alternative design principles.

Eliminate weak points

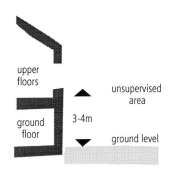

upper floors

unsupervised area

ground floor

3-4m

ground level

Eliminate weak points wherever the external envelope is not subject to natural surveillance.

If it is possible to approach parts of a building where there is no surveillance, the first 3-4 metres above ground level should be constructed without apertures, using materials with a strength similar to that of solid brickwork.

The same principle applies to higher parts of the building envelope that can be reached from flat roofs, fences, services pipework, etc.

Protect weak points

If there are any weak points in the building envelope, they should be strengthened or protected.

This applies if it is impossible to eliminate all apertures in the unsupervised parts of the building envelope. Doors or windows which can be reached from ground level should be of robust construction.

The use of bars, grilles or shutters should be considered.

Methods of protecting doors and windows are described in the existing security literature as well as manufacturers' data (see also 'Windows and burglary' and 'Security of doors' in Part 4, pages 149 and 153 respectively). Standard solutions tend to be recommended, but in practice there are many different options for the designer to consider. For example, locking ironmongery for opening windows is emphasised, but design options might include omitting opening lights and using alternative ventilation methods, or restricting the clear size of openings with closely spaced mullions or opening limiters.

It is often suggested that fire exit doors are a particularly weak point. Physically they are no more vulnerable to forced entry than other doors – less so if external ironmongery is omitted. Problems arise when they are used by building occupants as a routine short cut. Designers should plan fire exit doors so that they do not offer more convenient routes.

It is an advantage if fire doors can be monitored and supervised from inside the building either directly or by using alarms or CCTV.

If there are possible access points on the roof, any means of climbing onto the roof should be avoided.

Locks and shutters are a traditional strategy against common burglary, but all such devices can be overcome by determined attackers if they can work without risk of being seen and apprehended.

Enclose vulnerable areas

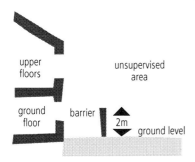

upper floors

unsupervised area

ground floor

barrier

2m

ground level

Prevent access to parts of a building which have weak points.

If there are weak points in the building envelope which are unsupervised and where heavy protection is undesirable or impractical, prevent access to the building face by enclosing the site with a boundary wall or fence at least 2m high, or with equivalent barriers such as hedging with thorny plants (as there is no natural surveillance there is no need for see-through barriers). Any gate needs to be of equal height and strength.

Footpaths and service roads to the side and rear should be avoided if possible. Such access will assist burglary on the unsupervised sides of the building.

The ideal arrangement is where one secure property backs onto another secure property.

Two further principles can also be considered in certain circumstances.

Alarm systems

Make provision for protected wiring runs for alarm systems.

Although alarm systems may not be regarded as part of inherently secure building design, as they require continual maintenance and monitoring, they can provide some additional protection by acting as a surrogate for the guardianship of occupants. Alarms are often required by insurers.

Alarm systems have developed a bad reputation due to the frequency of false alarms, but improved technologies provide opportunities to check or confirm the reliability of an alarm warning. For example, alarms combined with audio and video verification systems allow alarm monitoring stations to check why an alarm has been triggered.

To allow for the installation of alarm systems whenever required, non-residential buildings should be designed with suitable space for protected wiring runs around the perimeter of ground floors and protected telephone cables serving the building.

Internal secure areas

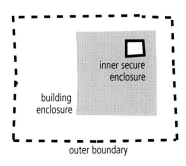

Provide an inner secure enclosure for high risk goods or equipment.

Where there are high risk targets in a building, such as money, valuable documents or goods, drugs, etc, these might be placed inside a further internal security enclosure within the building. The simplest version of this idea is a safe, but the same principle applies to secure store rooms or cages inside the building.

Where to find out more

Research sources

Bennett, T and Wright, R (1984)
Burglars on Burglary: prevention and the offender

Poyner, B and Webb, B (1991)
Crime Free Housing

Good practice

Association of Chief Police Officers (1992a)
Secured by Design: Commercial
This booklet is available from local police crime prevention services (ask for the Architectural Liaison Officer).

British Standard 8220: Part 2: 1987
Security of Buildings against Crime: Offices and Shops

British Standard 8220: Part 3: 1990
Security of Buildings against Crime: Warehouses and Distribution Units

■

2 SMASH AND GRAB RAIDS

Smash and grab raids are attacks on display windows or showcases. The raiders are not concerned about creating a noise or setting off an alarm. The essence of these attacks is surprise and a quick getaway.

The problem

There is little research on this well-known form of attack.

Officially smash and grab raids are classified as burglary, but the form of attack is distinctive. Unlike common burglary (§1), the essence is speed and surprise. The aim is to take a few high-value items and get away before there is time to respond to an alarm, or even for witnesses to intervene.

Smash and grab raids can be well planned attacks by experienced thieves or impulsive attacks by groups of rowdy youths or lone drunks. They usually occur when premises are closed and unoccupied, but have also been known to occur in busy times, such as an attack on a jeweller's window on a Saturday afternoon by a group of football supporters.

Much of the cost of smash and grab raids is in reinstating display windows. This can greatly exceed the value of stolen goods. Failed smash and grab raids can be indistinguishable from malicious damage or vandalism.

The cost of making good damage to the building fabric following a smash and grab raid is often greater than the value of goods stolen.

Where the risk occurs

The risk occurs primarily with shops displaying jewellery, consumer electrical goods and wine and spirits. However, any display window or showcase containing items of interest or value can be at risk. Display windows are particularly at risk in areas with poor surveillance, such as side alleyways, pedestrian subways or unstaffed transport facilities.

In general, the risk must be regarded as much less when the display windows or showcases are in an enclosed shopping mall, inside a shop or store, or within another building such as an office complex or hotel. In these more enclosed locations the potential attacker will find it more difficult to be certain of a quick getaway.

Approaches to prevention

There are several approaches that might be adopted to prevent smash and grab raids. One is to avoid displaying high-value goods in display windows, following the example, perhaps, of some building societies who fill their windows with three-dimensional advertising displays. Retailers might also consider substituting dummies for real goods, but few dummy displays can be made as attractive or compelling as the genuine product (and if they were too realistic they might attract raids).

Another approach is to tether display goods to prevent them being easily snatched from windows. However, if the tethering system is not obvious it may not deter potential thieves from smashing the windows.

The most practical and realistic approach seems to be to increase the strength or resistance to attack of the window, particularly where high-value and attractive consumer products are on display. The aim is to design the window or showcase to resist attacks from raiders armed with portable objects such as hammers, scaffold poles or bricks.

Design principles

Strengthen glazing

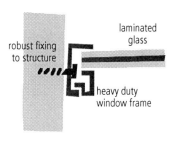

robust fixing
to structure

laminated
glass

heavy duty
window frame

Strengthen display windows.

Strengthening display windows is mainly appropriate in low or medium risk locations and where the contents on display are of moderate value.

Laminated glass is generally considered to be the best material. Specialist guidance can be obtained for its use in these applications.

Note that the frame into which the glass is fixed, and the fixing of the frame into the building structure, may need additional strengthening.

Insurers may not be content with this solution as the additional protection is not obvious to potential raiders. Glass may be damaged in an attempted raid, even if no goods are taken.

See also 'Windows and burglary' in Part 4, page 149.

Shutters and grilles

Protect particularly vulnerable display windows with shutters or grilles.

Shutters and grilles are intended for higher risk display windows, particularly where high-value goods are displayed, and also in higher risk neighbourhoods.

Shutters and grilles are often unpopular with planning authorities as they can dehumanize the appearance of shopping areas outside shopping hours. Solid external shutters are least attractive and can also attract graffiti (see §20). However, shutters do not need to be solid. There is a wide range of proprietary shutter and grille designs for use both outside and inside the glazing.

See also 'Windows and burglary' in Part 4, page 149.

Alternative window designs

Develop display window designs which avoid large areas of plate glass.

The ability of thieves to reach into a display to remove goods would be reduced by designs in which large areas of glass are replaced by small windows, rather like a modern version of the Regency bow window. Sub-division of the display into discrete units might also be considered. Some jewellery shops make a design feature of small viewing windows which highlight selected items.

Where to find out more

Good practice

Association of British Insurers (1991a)
Guidelines on Shop Front Protection
A short discussion of shutter and grille types, and possible conflicts with planning authorities.

British Standard 5544: 1978
Anti-bandit Glazing (glazing resistant to manual attack)

British Standard 8220: Part 2: 1987
Security of Buildings against Crime: Offices and Shops

Laminated Glass Information Centre (undated)
Glass & Architecture: Shopfront Security
A datasheet giving laminated glass thicknesses and maximum recommended window sizes.

Shop Front Security Campaign (1994)
Shop Front Security Report

■

3   RAMRAIDS

Ramraiding is an extreme version of smash and grab raids using a stolen vehicle to smash a way into the building. Despite its violence, it does seem to be a type of crime that can be readily controlled through design.

The problem

There is no published research available on ramraiding, but the phenomenon is well known. It is a form of burglary that has to be well organised because it requires several individuals to work together as a team. The nature of the attack is deliberately destructive and violent, and it is designed to be carried out quickly and with a flagrant disregard for any alarm systems.

Some of the attraction to the perpetrators must be the excitement of carrying out a daring raid. The whole operation has to be completed in a very few minutes, followed by a high-speed getaway.

The cost of making good the damage to building fabric and fittings following a ramraid is usually greater than the value of goods stolen. There may be justification for specific anti-ramraid protection to a building on the basis of life-cycle costing.

Where the risk occurs

Ramraiders are attracted by buildings containing very specific kinds of goods. Their main targets seem to be retail facilities containing consumer electrical goods like camcorders and computer games, spirits and cigarettes, and fashion and sports clothing.

The reason is that these goods are easy to re-sell and have a relatively high value by volume, which allows sufficiently rewarding quantities to be snatched, loaded and driven away quickly.

Many attacks are made on shops facing high streets. One daring and well publicised raid was on a consumer electrical store in an enclosed shopping centre. In this case vehicles first smashed through entrance doors to the mall before ramming the store entrance.

Further areas at risk are retail parks and industrial estates which are set back from main thoroughfares, providing opportunities for raiders to attack buildings without being seen or heard.

Approaches to prevention

Where retailers and others believe they are at risk from this form of attack, there may be opportunities for risk management. A first approach is to make ramraiding less rewarding by restricting the quantity of valuable goods in areas accessible to ramraiders, by moving some or all of the highly attractive goods on display into more secure stock rooms or security cages.

However, for many retailers there is a conflict between security and sales promotion, as there are compelling commercial reasons for sales displays to be abundantly stocked with attractive goods. In these situations, the most effective preventive approach is to use design to make ramraiding difficult or ineffective.

Design principles

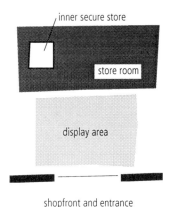

inner secure store

store room

display area

shopfront and entrance

Internal secure storage

Provide secure storage areas to reduce the quantity of target goods at risk.

The main function of secure internal structures is to slow down raids and reduce losses. Several 'layers', such as a cage within a store room, may increase effectiveness even further.

normal
construction

reinforced
construction

1m

ground level

Reinforced construction

Make ground level construction strong enough to resist the impact of vehicles.

Reinforced construction applies to ground floor walls, or walls around vehicle decks or podium areas which can be reached by vehicles.

There are no clear standards of resistance but solid masonary structure for the first metre or so is clearly better than lightweight cladding. Doors and windows are weak points. They can be protected with shutters, but raiders frequently breach conventional shutters.

In the case of shopfronts, the use of a 'stall riser' (a solid base to the shop window, about 600mm high) is better than full height glazing.

Restrict vehicle access

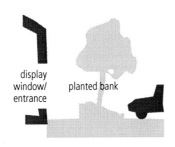

display window/entrance   planted bank

Design the areas in front of vulnerable points to prevent access or ramming manoeuvres by vehicles.

Preventing vehicle access to the building may be more effective than strengthening the fabric. However, the area that must be protected should allow for 'extensions', such as poles or iron pipes, that some ramraiders mount on vehicles.

There are many options open to the designer. The most basic is to install bollards (which can be removable), either in front of buildings or to prevent vehicle access to pedestrian areas. Other methods include the use of soft landscaping and changes of level, raised planters or other street furniture, and so on.

Access to vulnerable points such as loading bays can be prevented by providing fenced loading areas with lockable gates. This is particularly relevant in retail parks.

In general, the more indirect or complex the route that has to be taken by a ramming vehicle, the less attractive a target becomes for raiders. Their aim is for a speedy attack and the guarantee of an unhindered getaway.

Where to find out more

Good practice

Association of British Insurers (1991b)
Impact Attacks on Industrial and Commercial Premises using Vehicles

Shop Front Security Campaign (1994)
Shop Front Security Report

4   CASH ROBBERY

Wherever business transactions involve cash, there is a risk of staff being attacked or threatened with violence in attempts at robbery.

The problem

Cash is the most common objective of robbery in commercial settings. Although much robbery is aimed at cash while it is in transit, here we are concerned with attempts by robbers to force staff to hand over money from tills. All kinds of violence may be threatened but most often the threat is made with a weapon, such as a real or imitation gun.

It is normal practice to keep cash in some form of secure till at the point of transaction. This is largely to guard against petty theft and snatches. Tills function reasonably well, even though staff have to guard against theft if the till is left open and their attention is distracted. However, they do not protect against the risk of robbery. If staff are threatened by robbers most businesses advise non-resistance. Injuries to staff in control of tills can be serious and any attack is very frightening to the people involved.

Where the risk occurs

The risk can occur wherever cash is held. In practice, the risk occurs most frequently at the point where money transactions between staff and the public take place.

In theory the level of risk will be highest where there are large amounts of cash, but current practice means that these situations are often well protected; there are greater risks where there are more modest amounts of money and less protection.

The risk of robbery can be expected where cash security is a less overriding consideration than providing a quick and economic service to customers. Research from Britain and the United States suggests that the risk is greatest in shops or convenience stores which are small or stay open late and so have only one or two members of staff on the premises. Also at risk are cash tills operated at quiet times by very few staff, for example in railway ticket offices and car park pay kiosks.

Approaches to prevention

There are several strategies which can be used alone or in combination to reduce the risk of robbery. The three basic ways of protecting staff are by target hardening, target reduction, or target removal.

For robbery, target hardening means protecting staff, normally by permanent glass screens or rising screens. Some protection is provided by other staff, particularly in low risk situations like retailing or leisure facilities.

Target reduction means reducing the cash held at the point of transaction, by holding it elsewhere and passing it to a bank or security courier as quickly as possible.

Target removal means either using cashless transactions or automating cash transactions to avoid the need for staff. Both of these options are now well developed.

Many cash transactions now take place using automatic machines, like the familiar cash dispenser.

There are also some more general design strategies to deter potential robbers. The context is critical. For example, a till located at the back of a well staffed retail store may be less vulnerable to robbery than one which is close to a door; whereas in a small convenience store open at night with only one member of staff, it may be safer to place the cash till at the front of the store in clear view of a busy street.

A good deal of research has been done in the United States on using design to reduce night-time robberies at convenience stores (summarised in Hunter and Jeffrey, 1992). Design features which seemed important included: reducing opportunities for concealment at entrances by lighting and location, making the interior more visible from the street, lighting adjacent car parking areas, and locating stores near to other evening activities.

Further evidence that design and layout affects the risk of robbery comes from an American study of banks and building societies (Wise and Wise, 1985). Researchers found that the wider the spacing of till positions and the more spread out the layout of the counters, the less the risk of violent attack. Robberies occurred more often when the customer area was compact and the robber(s) could control the whole area visually. The study also found that premises with only one public entrance had a higher risk of robbery, presumably for the same reason.

Design principles

Storage for surplus cash

shopfront and entrance

Provide secure cash storage away from the point of transaction.

To minimise the amount of cash in tills, surplus cash should be stored elsewhere than at the point of transaction. Provide a safe in a secure room out of sight of areas normally accessible to the public, the more secret and inaccessible from public areas the better.

Security screens

Protect staff by security screens.

Where there are large amounts of cash and only one or two staff present, there is good evidence to show that strong physical resistance is required to protect staff. Examples include smaller post offices (Ekblom, 1987), ticket offices and pay kiosks at car parks.

Public and staff areas should be separated by a continuous counter, fitted with a security screen. Screens need to give protection against the use of firearms and chemical sprays by ensuring that there is no straight line of fire from the public side of a counter to the staff side. Also screens must resist attacks by sledgehammers and prevent access over the top. The use of high specification laminated glass is considered essential.

This arrangement is commonly used if cash transactions are for documents rather than goods, as in banks, building societies, post offices or ticket offices; it not used in retailing.

Where cash is handled in isolated parts of a structure, such as kiosks or ticket offices, the protection must extend to the whole structure. For example, the door will need to be at least as strong as the glazed screen. Ideally places where cash is handled should not be isolated, but integrated into a larger structure for increased security.

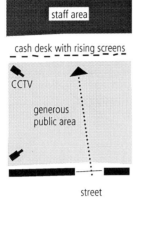

cash desk with rising screens

CCTV

generous
public area

street

Rising security screens

Protect staff with rising screens that are activated in emergencies only.

Rising screens clearly provide less security than fixed screens, but where several staff (six to eight or more) are working together and where a good public image is of particular importance, for example in building societies, they may be acceptable. Rising screens are usually fitted in combination with low fixed screens, typically in perspex.

Where the reduced level of security given by rising screens is accepted, it is important to create a quiet and spacious public area with a good distance between the entrance and the counter. This gives the staff a better opportunity to recognise trouble in time to activate the screens. It is also common to cover the public area with CCTV surveillance to record any incident and assist detection. Under such conditions CCTV may act as a deterrent.

Some organisations control access to this outer area; customers have to wait outside glazed entrance doors until staff release the doors by remote control.

Visibility of cash tills

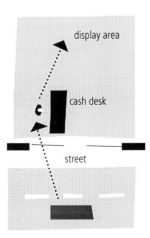

display area

cash desk

street

Ensure that cash tills are highly visible from inside and outside the premises.

With modest amounts of cash and only one or two members of staff, the risk of robbery is mainly due to the isolation of staff. Attacks occur most often in the evening when business is quiet, in settings such as convenience stores, off licences and filling stations.

Research, mainly from the United States, suggests that a range of measures can protect staff and reduce the incidence of robbery. They include making the cash till more visible from the street by keeping windows clear, lighting it well and putting it at a raised level. Raising the till may make attacks less easy and allow staff to get a better view of the store. Lighting the areas outside may discourage potential robbers by increasing the chance that they would be seen and interrupted.

Note that at night many filling stations abandon the idea of providing a convenience store service; payments for fuel are made at a till protected with screens.

Grouping of cash tills

Group several cash points together.

Where modest amounts of cash are held and there are many staff on the premises, the risk of robbery is much lower. Good practices have been developed which can be expected to enhance security. In particular, the practice of grouping several tills together in large open plan stores and supermarkets reduces their attractiveness as a target for intended robbers.

Where to find out more

Research sources

Austin, C (1988)
The Prevention of Robbery at Building Society Branches

Ekblom, B (1987)
Preventing Robberies at Sub-Post Offices: an evaluation of a security initiative

Hunter, R D and Jeffrey, C R (1992)
'Preventing convenience store robbery through environmental design'

Wise, J A and Wise, B K (1985)
'The interior design of banks and the psychological deterrence of bank robberies'

Good Practice

British Standard 5051: Part 1: 1973
Bullet-resistant Glazing

■

5 VIOLENCE TO STAFF

The problem of protecting employees from violence other than robbery. It is
is predominantly a problem of verbal abuse, but physical violence does occur.

The problem

Although violence or threats of violence associated with the handling of money are recognised as a common risk at work (see §4 'Cash robbery'), it is now also recognised that violence for other reasons is an even greater problem. In practice, the violence is more likely to be verbal rather than physical, but it is clear that staff in many forms of employment consider verbal abuse and the threat of violence to be a serious issue. Examples include all those in contact with the public in transport, health, education, social services, retailing and entertainment – notably pubs and clubs.

A recent survey of retail crime reported that about eight staff per 1000 experience threats or violence from the public each year (Burrows and Speed, 1994).

Where the risk occurs

Essentially, staff are most likely to be in abusive or violent confrontation when controlling or checking on the behaviour of people, such as checking tickets on transport systems or restricting access to sporting or entertainment facilities. Situations which often give rise to violent behaviour are government offices dealing with social welfare benefits or the allocation of housing, and social work and health care facilities. Other well known settings include public houses and, at a lower level of risk, almost any entry and reception point (for further examples see Poyner and Warne, 1988).

Approaches to prevention

The control and management of this risk involves a very wide range of measures, including workplace design, procedures for dealing with the public, and staff training in awareness and interpersonal skills. Design can contribute only a part of any preventive policy, but where potentially violent or abusive interactions take place in a predictable location, there is quite a lot that a well-designed environment can contribute.

It is unrealistic in this guide to cover the whole range of possible settings in which design can help. What follows are some principles for the design of enquiry and reception counters and interview areas.

Design principles

The first two principles apply to the design of counters for transactions such as enquiries, reception, registration, serving drinks, etc. These may be in entrances, reception areas, information centres or even bars.

Counter height and depth

preferred face-to-face distance 1500mm

800mm min

counter

1200mm max

Make counters high and deep enough to separate staff and customers.

It is important to reduce the risk of abusive customers reaching over a counter or intimidating staff. Avoid low and narrow counter designs. The actual dimensions of a counter are determined by the requirements of the transaction but counters under a metre high on the customers' side are too low. Heights up to 1200mm should be considered. Counter depths should be at least 800mm.

A two level counter design is often

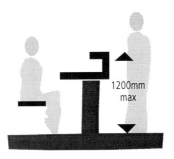

used, for example in hotel reception counters. This gives greater security to the staff side of the counter.

With higher and wider counters, face-to-face distances increase to around 1500mm which is likely to make abusive behaviour less frequent and easier for staff to cope with.

Raised floor

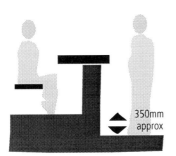

Use raised floor levels where members of staff are seated.

If staff remain seated at counters while customers are standing, there is some advantage in raising the floor level on the staff side so that eye contact is at the same level on both sides of the counter.

The following principle applies to the design of interview areas.

Wide tables and low chairs

Provide wide tables and low chairs to separate staff and interviewees.

If the interview is across a table, then it is important to encourage face-to-face distances of about 1500mm, even if the interviewee leans on one side of the table. Low chairs also contribute to greater face-to-face distances.

The following principles apply to the design of both counters and interview areas.

Escape route

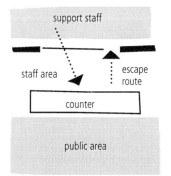

Provide an escape route for members of staff.	In planning the staff area behind a counter it should be possible for staff to have direct access to a door or other opening to an area where supporting staff will normally be on hand.
	A similar escape route should be provided in interview settings.

Support surveillance

Provide support surveillance from other members of staff.	Where possible a counter should be overlooked from areas where other staff are normally at work, such as a neighbouring office. This might be achieved by the use of a glazed wall to interview areas. Isolated counter staff are naturally more at risk.
	Panic buttons are often suggested at counters or in interview situations, but they are rarely used.

Where to find out more

Research sources

Burrows, J and Speed, M (1994)
Retail Crime Costs 1992/93

Health and Safety Commission (1987)
Violence to Staff in the Health Services

Mayhew, P, Elliot, D and Dowds, L (1989)
The 1988 British Crime Survey
See chapter 4 'Crime at work'

Poyner, B and Warne, C (1988)
Preventing Violence to Staff

6 SHOPLIFTING

Unlike other security issues, shoplifting is by definition specific to one type of environment – shops and retail stores. It involves the theft of displayed goods by customers or individuals posing as customers.

The problem

Almost any small portable item of merchandise can be stolen from retail displays. Most of the items taken represent everyday purchases. In a study of shoplifting in London's Oxford Street (Poyner and Woodall, 1987), two-thirds of the items recorded were clothing. They included almost everything from socks and tights to trousers and jumpers and every kind of underwear. Other items included household goods, cosmetics and entertainment goods like music and video cassettes.

Although this is a well recognised type of crime, knowledge of its extent is obscured by statistical problems. It is difficult for retailers to distinguish between shoplifting and other stock losses. Theft by staff, fraudulent delivery practices, damage to goods and legitimate customer exchanges or 'deals' make it difficult to estimate exact losses.

Researchers have found remarkably high rates of theft among shoppers. For example, Buckle and Farrington (1984) found from observation of customers in a department store that nearly two in every hundred took items from displays without paying for them. Even higher rates have been reported in the United States. These observations have been supported by more detailed counting of small items displayed and sold in electrical chain stores. Here it was found that on average 11% of some small items removed from displays were stolen (Buckle et al, 1992).

Where the risk occurs

In modern retailing, goods are highly accessible to customers, and therefore to shoplifters as well.

Common sense suggests that the extent of shoplifting will vary according to the type of shop. It is unlikely to be a problem in stores such as furniture or carpet stores where most items cannot be readily concealed on the person. But wherever there are easily accessible displays of goods and particularly where there are low levels of staffing, there is a risk of shoplifting.

The Oxford Street study showed that most arrests for shoplifting occurred in open plan stores where goods were on self-selection displays. Shops like jewellers and shoe shops appeared to have few problems of shoplifting because customers were served directly by shop staff and only allowed to handle goods under their direct supervision.

Approaches to prevention

It is clear from these research findings that one way to reduce shop theft would be to return to more traditional methods of service, with goods presented across the counter or with intensive personal service, such as in a small dress shop. However, this approach is generally considered unrealistic in the modern world of retailing where direct customer access to attractive displays both increases sales and reduces staffing costs.

A modern development which combines the security of traditional methods with attractive displays is the catalogue purchase shop in which customers select their purchase from a catalogue or window display and receive the goods at the counter.

If goods are to be openly displayed for selection by customers, then there are a number of ways of reducing theft. One is the use of devices to secure goods on display.

For example, security loops or chains may be used when displaying valuable electrical goods, such as radios or CD players, and also expensive clothing. Also, there are 'electronic article surveillance' tagging systems designed to trigger an alarm if goods are taken through an exit before the tag is removed.

There is research evidence that security chains can be effective when used with more exclusive fashion clothing (Poyner and Woodall, 1987). Alarmed loops and tagging systems may also be effective (Farrington et al, 1993), but they are plagued with false alarms and most tags are readily removed or defeated.

Another technique used by retailers is to make the potential target of theft more difficult to conceal, for example by mounting small items in bubble packs on display cards. The overall size of the card can be several times the size of the merchandise.

Perhaps the most acceptable approach to theft control for self-selection displays is surveillance by staff. Whether the retailer relies solely on sales staff or employs specialist security staff, the design and layout of retail space can greatly affect their ability to watch over displayed goods.

Although the value of design as an aid to surveillance is not well researched, there is evidence from a study of electrical retail stores that when stock around the staff and payment counters was rearranged to improve surveillance, theft was significantly reduced. This rearrangement appeared to be more effective than the employment of a security guard (Farrington et al, 1993).

CCTV appears to be a useful means of surveillance and may well act as a deterrent to some degree. However, its effectiveness depends on how it is deployed. The evidence for its value in controlling shoplifting is not well established in research. Doubts are clearly illustrated in a study of a large tape and record store by Ekblom (1986). He commented that, "...none of the offenders arrested in the study period was caught using the CCTV"; and this was in a store with one of the highest arrest rates in the country.

Design principles

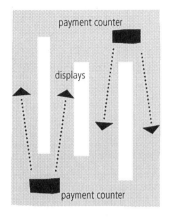

Maximise the surveillance of sales floors

The layout of self-selection retail floors should maximise the surveillance of display units by staff, particularly from service or pay points.

Layouts with high level partitioning and displays which block surveillance should be avoided. Aisles or gangways should be aligned to assist surveillance from staff service and pay points.

The use of open wells or atria in multi-level shops allows surveillance from upper levels or from stairs and escalators rising through the store.

Avoid congested layouts

The width of main aisles between self-selection display units should generally be three metres or more.

Congested aisles or gangways should be avoided as they make surveillance by staff impossible.

It was found in a study of markets in Birmingham that when the width of gangways between stalls was increased from 2m to 3m there was much less theft from shopping bags (Poyner, 1983). Although the study was not measuring shoplifting, it seems reasonable to assume that wider gangways also discourage shoplifting by increasing the opportunity for surveillance by staff and other shoppers.

Minimise the number of exits

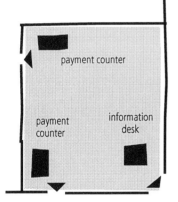

Minimise the number of exits from retail premises, and locate permanently staffed work positions close to every exit.

To improve the deterrent effect of surveillance, the design of space around entrances is important because shoplifters know they are at risk of being apprehended as they leave the premises.

Supervision of exits by staff is more difficult if there are many entrances and exits. By reducing the number it should be possible to arrange activities and staffing so that all exits are supervised. For example, placing information points, special displays or simply well-staffed counters near the exits would be good practice.

Where to find out more

Research sources

Buckle, A and Farrington, D P (1984)
'An observational study of shoplifting'

Buckle, A, Farrington, D P, Burrows, J, Speed, M and Burns-Howell, A (1992)
'Measuring shoplifting by repeated systematic counting'

Ekblom, P (1986)
The Prevention of Shop Theft: an approach through crime analysis

Farrington, D P, Bowen, S, Buckle, A, Burns-Howell, A,
Burrows, J and Speed, M (1993)
'An experiment on the prevention of shoplifting'

Poyner, B (1983)
Design against Crime

Poyner, B and Woodall, R (1987)
Preventing Shoplifting: a study in Oxford Street

Good practice

British Standard 8220: Part 2: 1987
Security of Buildings against Crime: Offices and Shops

Jones, P H (1990)
Retail Loss Control

■

7 EMPLOYEE PILFERAGE

Unfortunately, not all employees can be trusted with the goods and supplies which they must handle during their work. Although most control systems have little to do with building design, there are some design principles which may help.

The problem

The level of risk depends on the value and portability of goods and supplies in the workplace and on management practices. The extent of employee theft is almost impossible to establish in many businesses. For example, retail losses through employee theft are hard to separate from shoplifting, and in production processes it is often difficult to distinguish between theft and wastage, such as the wastage of food in the hotel and catering industry.

Where the risk occurs

Most retailing and manufacturing processes which involve portable and readily usable or resaleable goods and materials are at risk. This includes most shops, warehouses and workshops, as well as premises such as hotels and hospitals.

Approaches to prevention

It is probably true that for this security issue non-design methods of prevention are the most important. Much depends on supervision and on the motivation of staff to be honest through the use of accounting and stock control systems and appropriate (dis)incentives.

Nevertheless, there are some aspects of building design and layout which can assist these managerial controls. The following design principles may apply to various employment situations. Any application and development of these design approaches should be discussed with management in the early stages of a design project.

Design principles

Hiding places

Avoid opportunities for concealing goods for retrieval later.

Any yards or external areas which are unlocked when the premises are closed should be free from hidden recesses, storage bunkers, refuse containers and clutter to prevent the concealment of goods which can be retrieved outside working hours. Ideally, such areas should be avoided wherever possible, or kept within a secure external boundary.

Inside the building, the location of storage areas for staff personal belongings should not be readily acccessible during routine work to avoid concealment of stolen property.

Supervised staff exits

Exits for staff should be readily supervisable.

In some cases exits may be supervised by door- or gate-keepers. Car parking should also be designed to be supervised either formally or informally (by overlooking from offices or similar accommodation).

Staff should never enter or leave the premises through a loading or storage area. Car parks should be completely separate from loading areas and inaccessible (even by pedestrians) from loading areas.

Access via changing areas

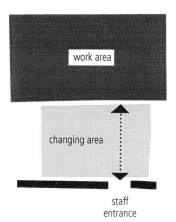

Access/exit to work areas only via changing facilities.

Where businesses consider the policy appropriate, staff access to work areas might be through supervised changing areas with personal lockers. Consideration might also be given to the use of lockers with transparent cage construction.

Separation of storage areas

Access to stores and warehouses restricted to authorised personnel only.

The building layout should avoid any circulation routes passing through storage areas.

The number of access points to storage areas should be minimised. Access points can be controlled by devices such as coded locks or electronic pass cards. It may be a useful additional safeguard if access points can be easily overlooked from supervisory or management offices.

High risk goods

Provide segregation of high risk/value items within storage areas.

Various methods can be used to segregate high risk goods, for example, storage cages within stores which can be supervised from outside but only accessed by locks, coded keys, etc.

Where to find out more

Good practice

Barefoot, J K (1990)
Employee Theft Protection

■

8 CUSTOMER BELONGINGS

Wherever members of the public are receiving a service and have to be temporarily separated from their belongings, there is a need to protect their property from theft or damage.

The problem

There are many situations in which people need to leave their personal belongings. The most common is the removal of clothing. Outer winter clothing is nearly always taken off once people have entered a warm environment to stay for some time, such as in an office or other workplace or a restaurant. More extensive removal or changing of clothing is necessary for sporting and recreational activities as well as medical examination and treatment. In all these situations there is always some risk of theft of clothing, but more particularly of the theft of valuables from wallets and handbags, etc. The potential thief might be a member of staff, another customer, or anyone with access to the facility.

Although the level of theft is not well documented, there is good reason to believe that the problem needs more consideration in design. There is evidence to show that many individuals are concerned about the problem and reluctant to leave their belongings where they cannot guard them effectively.

Where the risk occurs

Each problem needs to be considered carefully as risks vary, but the following examples illustrate commonly recognised situations.

- Hats, coats, bags, etc, in restaurants, hairdressers, offices, theatres, clubs, etc

- Fitting rooms in clothing shops

- Changing rooms in sports and leisure facilities

- Changing for medical examination/treatment

- Clothing and valuables left in hotels

- Temporary storage of clothing and baggage during travel (bus, rail and air interchanges).

Approaches to prevention

There are a number of conventional solutions to this problem which are found in specific building types, for example, cloakroom attendants at theatres, personal lockers at swimming baths, coat hooks in offices. These standard solutions tend to obscure the underlying strategies and principles.

Managements of buildings must decide how far they take responsibility for customers' belongings. Much depends on the activities and circumstances of the customer.

There are two basic approaches.

First, where customers are mobile and physically active, such as playing a sport, dancing or simply moving about at a social event, management will need to take a direct responsibility for providing secure storage. This also applies in situations where customers are not free to take responsibility for themselves because they are receiving treatment – at a hairdresser, in a dentist's chair, on a massage couch, undergoing a medical procedure, etc.

All Things that cannot be always kept lockt up in Store-rooms, ought however to be kept in some Place where they may be constantly in Sight; and especially such things as are seldomest in Use; because those things which are most in Sight, are least in Danger of Thieves.

Alberti *The Ten Books of Architecture* (1472)

Second, where they are likely to remain in one location, for example in a restaurant, customers are capable of keeping an eye on their own possessions.

Design principles

The design principles can be grouped under the two basic approaches.

It is worth noting that there are many well established practices where the principles are not fully adhered to, which might be considered more carefully. For example coat storage in restaurants may be out of sight of the customers but not fully supervised by the restaurant staff, or changing cubicles in hospital out-patients departments may be poorly supervised by staff while patients are undergoing treatment.

Supervised storage

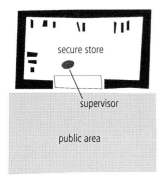

Provide a secure area in which property can be stored under the watchful eye of a member of staff.	Examples of supervised storage include a cloakroom with an attendant, a hotel baggage room supervised by a porter, or a left luggage office.

Lockable secure storage

Provide secure, lockable storage to which only the customer has access (by key or code).	Examples of lockable secure storage include personal lockers in sports changing rooms, left luggage lockers and personal safeboxes for hotel guests' valuables.

For situations where customers take responsibility, there is one basic principle.

Storage supervisable by the customer

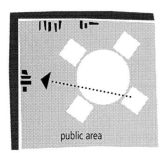

public area

Provide storage devices or equipment in a place that can be easily supervised by the customer.

Storage supervisable by the customer might include a coat rack or cupboard close to a restaurant table or in an office, conference or meeting area. It might be a changing cubicle in a clothing shop that can be supervised from the selling floor, or one which forms an integral part of a medical examination room.

Where to find out more

This is a security issue which appears to have received little attention from research or published design guidance.

■

9 PERSONAL BELONGINGS AT WORK

The problem of providing staff with safe places to keep clothing, shopping, other personal belongings, and especially valuables such as wallets and purses, during working hours.

The problem

The British Crime Survey began to ask about crime victimisation at work in 1988. Unpublished figures from the Survey indicate that about two percent of employees experience personal theft at work each year. For some reason men are more often victimised than women. Perhaps this is because men frequently leave wallets in jackets which they take off while working, whereas women may look after their handbags more assiduously. However, there are many workplaces in which staff have little or no opportunity to keep their valuables in a safe place.

Where the risk occurs

Although there is little research, the most common potential thieves are believed to be outsiders and visitors to workplaces rather than other staff. The problem is unlikely to be serious in small work groups where personal belongings can be well supervised. It is likely to be most severe in large buildings with easy access for non-staff, such as some public offices and buildings like hospitals.

Approaches to prevention

Broadly, there are two approaches to prevention. The first is to emphasise the need for staff to take care of their possessions and not leave them unguarded. A more realistic approach is for employers to take personal security more seriously and provide staff with adequate lockable storage for personal possessions while at work.

Note there is a risk that personal storage may be used to conceal the employer's goods (see §7 'Employee pilferage').

Design principles

There seem to be two alternatives:

Personal lockers

Provide lockers for personal storage.

These can be provided close to the workplace or in supervised locker rooms (see §7).

A common location which provides some general supervision of lockers is in well used corridor or circulation spaces. This may require approval from the fire officer.

Storage at the workplace

Provide lockable drawers or cupboards as part of the furniture and equipment at the workplace/ workstation.

For example, a lockable desk drawer large enough to keep a handbag and a small amount of shopping.

Where to find out more

There does not appear to be any research or design guidance on this aspect of inherent security.

10 OTHER THEFTS FROM PREMISES

Wherever buildings are accessible to the public there is a risk of portable materials, supplies, furnishings, etc, being stolen.

The problem

Much petty theft from buildings is related to everyday objects, such as ashtrays, toilet paper, soap, cutlery, etc, but almost every usable or valuable item that is easily portable is at risk. Most organisations could report the loss of significant items, such as furniture, office equipment, works of art and other decorative items.

Where the risk occurs

Buildings, such as hotels, hospitals and educational establishments, have much equipment and furniture which can be removed and is often stolen. There are also public areas of buildings which are particularly at risk, such as waiting areas and toilets.

Approaches to prevention

The problem is generally well understood by building managers. There are a number of commonly used approaches to prevention. They can be classified broadly as risk management, target hardening and supervision – the latter two being of concern in design for inherent security.

It is common practice for businesses, such as hotels, restaurants, airlines, local government, etc, to mark goods and supplies so that items of china, cutlery, towels, ashtrays, etc, do not have a significant market elsewhere.

The use of disposable items like towels, cups, plates, cultery, etc, is mainly designed to reduce laundering and washing-up costs, but it also eliminates some aspects of theft.

Where items are unavoidably at risk, it is prudent to adopt low cost specifications, for example, using cheap tungsten light bulbs in hotel bedrooms rather than more expensive low energy lamps. This reduces both the likelihood and the cost of theft.

Another approach is to make potential targets for theft less portable. Items of furniture and equipment in public areas (particularly unsupervised areas such as toilets and guest bedrooms in hotels) should be fixed or made too large or heavy to be removed easily.

Supervision by staff is also an important means of reducing theft. Wherever possible, public areas should be open enough to allow at least some casual surveillance by staff and/or other customers, and exits should be subject to surveillance.

Design principles

The decision about which of these approaches is most appropriate must be made by the building management. If this decision is to employ target hardening and staff supervision, there are implications for design.

Target hardening

Replace portable items with alternatives that are fixed or too large or heavy to be removed easily.

Examples of substituting fixed for portable equipment or materials include the use of roller-towels, hot air hand driers, soap dispensers, built-in light fittings.

Where possible, fix items of furniture and equipment to the structure – for example, use fixed seating and screw pictures to walls.

Use only large and heavy furnishing, for example, large planters or weighted litter bins and ashtrays. This approach also applies to landscape elements externally.

Supervision

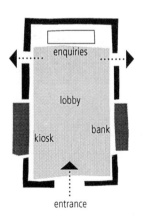

Design public areas to maximise supervision by staff and/or customers.

This is perhaps the most effective method of preventing theft, but clearly there are limits to its practicality. For example, a high class hotel will benefit from relatively high levels of staffing to maintain the security of hotel property, but such levels of staffing will not be justified in many other settings.

Surveillance of exits

Reduce the number of public exits from buildings and provide surveillance.

Wherever possible, reduce the number of public exits from buildings (including fire exits) and provide surveillance. This might require door or reception staff, but may be achieved more informally by surveillance from showroom staff, information desks, or staff in shops or cafes which open off an entrance foyer, and so on.

Note that there are obvious similarities to §6 'Shoplifting'.

Where to find out more

There does not appear to be any research or design guidance on this aspect of inherent security.

■

11 PARKED CARS

Whenever cars are left unattended in car parks there is a risk that they might be stolen, broken into, stripped of items such as wheels, or simply damaged in some malicious way.

The problem

Car crime is a major problem in Britain (Webb and Laycock, 1992b). Roughly two in every 100 cars are stolen each year. Reported thefts *from* cars have now become about twice as common as the theft *of* cars and it is widely recognised that many such incidents go unreported. Recent figures from the 1994 British Crime Survey suggest that whereas 92% of thefts of cars are recorded by the police, only 31% of thefts from motor vehicles are recorded (Mayhew et al, 1994). Criminal damage to vehicles appears to be a less serious problem but information on this is much less clear.

The British Crime Survey has estimated that about 20% of car crime occurs in public and private car parks, but another study (Webb et al, 1992) showed that in some police force areas the proportion was twice as high, with 40% of car crime occurring in car parks. This study also showed that the level of crime varied widely. Some railway commuter car parks had rates of over ten cars stolen per 100 parking spaces per year, and several car parks had double this rate for thefts from cars.

Car parks at commuter railway stations have some of the highest reported levels of car crime.

CIRIA SPECIAL PUBLICATION 115 · DESIGN FOR INHERENT SECURITY · 1995

Where the risk occurs

The risk varies according to the type of car, the area in which it is parked and the type of parking facility. Some makes and models of cars are more frequently stolen than others (Clarke and Harris, 1992; Houghton, 1992).

Cars are perhaps most at risk when parked on open parking lots surrounded by waste ground, or in run-down urban areas with many vacant or derelict properties. Safe places to park would be small, well-supervised staff parking areas at the back of small office or shop premises, or well-supervised open commercial car parks with controlled entry and exits. Long-stay car parks, commuter parking at railway stations or all-day parking in commercial districts tend to carry higher risks than short-stay parking, for example for shopping.

Approaches to prevention

The most obvious approach to prevention of theft is the use of security devices on cars. The provision of door locks has been standard practice on cars since the early 1930s; the key operated ignition/starter switch was introduced in America in 1949; and by 1969 steering column locks were required by law in the UK (Webb and Laycock, 1992b). Continuing concern over vehicle crime has led more recently to a further voluntary increase in the specification of security devices by many car manufacturers, and there is a thriving industry in add-on devices.

In 1928 the Minister of Transport passed an order making it illegal for drivers in London to lock their cars when parked in public places... The rapid growth of car ownership ... meant that obstructions caused by parked cars were a serious problem, and it had to be possible to move cars by hand in the absence of their owners.

Webb and Laycock (1992b)

Although the evidence for crime prevention by improved in-car security appears to be encouraging (Houghton, 1992), there is also a good case to be made for environmental design as another way of reducing the risk of car crime. Two studies (Poyner, 1992; Webb et al, 1992) provide strong evidence to show that car park design and parking systems greatly effect the risk of theft. Unlike in-car security measures, this approach can also reduce the risk of criminal damage to cars.

Both of these studies show that the most effective measures to prevent the theft of cars from car parks are

Research has shown that ticket-controlled access barriers are effective in reducing the theft of cars from car parks.

close supervision or control of the exit. Ticket-controlled exit barriers are particularly effective. Supervision of exits by car parking staff or even by associated business activities, such as a 24-hour taxi service, can virtually eliminate the illegal removal of cars.

Preventing theft from cars is more problematic. Supervision of exits is not enough. The aim must be to discourage potential thieves hanging around parking areas and to provide as much surveillance as possible. Pedestrian access must be limited so that only those who are clearly car users have a legitimate reason for being in the parking area. Staff supervision reduces the risk, and in short-stay parking the frequent movement of legitimate users may provide a sufficient deterrent. In long-stay commuter car parking or overnight parking greater supervision is needed, possibly with the aid of CCTV.

Note that managed parking, where cars are handed over to parking staff or placed in automatic parking systems, provide even more security.

Design principles

The following three principles apply to all self-parking arrangements.

Control exits from parking

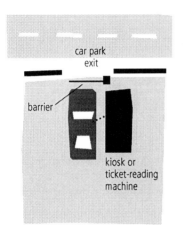

car park exit

barrier

kiosk or ticket-reading machine

Exits from car parks should be controlled by staff supervision or an automatic ticket system.

In public car parks where there is some form of charging system, the method of control should apply at the exit, either by paying an attendant or operating a ticket-controlled barrier. Cars are much more likely to be stolen from public car parks without ticket or staff controls at the exit.

The risk of cars being stolen is less in busy short-stay parking, such as at local shopping centres or supermarkets.

In the case of private car parks, exits should be designed to be supervised by reception or security staff. It may be practical in some settings to control entry and exit by a lockable gate or barrier operated by key- or card-holders only.

Reduce pedestrian access

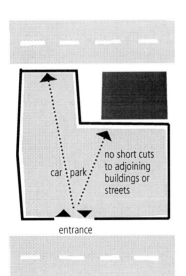

no short cuts to adjoining buildings or streets

car park

entrance

Public pedestrian access to car parking should be limited to one or two points; there should be no reason for non-users to walk through or take short cuts.

The objective is to reduce the opportunity for potential offenders to hang around a car park looking for possible targets. It requires car parks to be enclosed with fencing or similar barriers. This applies equally to an open car park and the ground level of a multi-storey car park.

In the case of some private car parks or customer parking for retail facilities, it should be possible to channel pedestrian access to one or two supervised entry and exit points. In this way non-legitimate users would be easily observed entering the car park.

Surveillance of car parking

Maximise surveillance by parking attendants, security staff and legitimate users.

In practical terms this means that the car park layout should be based on easily observed rows of cars without small pockets of parking obscured by service cores, columns, etc.

Surveillance also comes from users driving around the car park, so it helps security if the flow of cars moves through the whole parking area.

Public car parks benefit from the presence of offices for security or parking staff, or businesses such as taxi or car hire, which provide natural surveillance. In short-stay parking, open layouts with strategically located entrances or lift lobbies can provide natural surveillance from the continuous flow of legitimate users.

The 'pay-on-foot' system used in some public car parks seems to have advantages for surveillance, in addition to providing ticket-controlled vehicle exits (see design principle 'Control exits from parking'). Since pay machines are normally provided at pedestrian entrances on each floor or parking area, such systems require that the staff who service the machines and deal with problems move frequently from floor to floor.

Private car parks benefit from being overlooked by windows in the premises they serve, such as offices or factories.

Surveillance is much more difficult for long-stay car parks, such as all-day commuter parking and overnight parking. Here security staff, perhaps supported by CCTV, may be essential to prevent car crime. Serious consideration might be given to managed parking systems rather than self-parking. Cars that are taken by

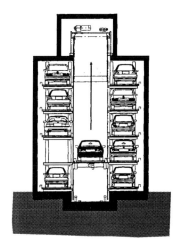

car park staff to secure pounds or block-parked will not be accessible to potential offenders.

The most secure option may be automated parking or stacking systems where cars are moved and parked on trolleys. However, these are expensive and not convenient for short-stay usage.

The lighting of car parks is essential for surveillance and for reducing fear and anxiety. Good quality of light and an even distribution are more important than the brightness of car park lighting (see 'Lighting and security' in Part 4, page 158).

Where to find out more

Research sources

Clarke, R V and Harris, P M (1992)
'A rational choice perspective on the targets of automobile theft'

Houghton, G (1992)
Car Theft in England and Wales: the Home Office Car Theft Index

Mayhew, P, Mirrlees-Black, C and Aye Maung, N (1994)
'Trends in Crime: findings from the 1994 British Crime Survey'

Poyner, B (1992)
'Situational crime prevention in two parking facilities'

Webb, B and Laycock, G (1992b)
Tackling Car Crime: the nature and extent of the problem

Webb, B, Brown, B, and Bennett, K (1992)
Preventing Car Crime in Car Parks

Good practice

Association of Chief Police Officers (1992b)
Secured Car Parks

Two checklists for evaluating outdoor and enclosed car parks, highlighting features which are expected to reduce car crime and other crimes in car parks (see 'Public car parks' in Part 3, page 134).

■

12   BICYCLE PARKING

This is the problem of providing safe places to leave bicycles. Although many environmental policies (to reduce pollution and transport congestion) favour the use of bicycles, fear of theft discourages their use.

The problem

Official statistics record that over 200,000 bicycles are reported missing to the police in England and Wales each year, and surveys indicate that many cycle thefts are unreported (Wheeler, 1989). It is clear that cyclists are concerned about security and that fear of theft is a significant factor which inhibits the use of bicycles (McClintock, 1992).

Where the risk occurs

Because of a shortage of safe places to leave bicycles, cyclists exploit any available fixture.

There is little research to give a more detailed picture of the problem or to indicate the levels of risk involved in different locations. The risk is present at non-residential buildings wherever bicycles are used for journeys to work, school, shopping, transport facilities, etc.

Areas such as shopping centres, schools and university campuses are certainly known to have problems. All-day bicycle parking for commuters at railway stations is one of the most challenging problems.

Cycles may often be 'borrowed' for a single journey and then abandoned, rather than being kept or re-sold by the thief.

Approaches to prevention

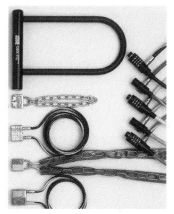

Locks are a vital part of bicycle security, but good surveillance is important to prevent locks being forced.

As in other security issues, the two basic approaches to preventing theft are by locking the property at risk or by guarding it, or a combination of both.

Most bicycle users recognise the need to lock the wheels of their machines to the frame when left unattended, so that they cannot simply be ridden away. To be more safely secured so that they cannot be carried away, bicycles should also be locked onto some fixed object, such as a cycle stand, railing or other piece of street furniture.

Whilst bicycles can be protected more securely with high quality locks there is only limited protection from basic cycle locks. For this reason, good security for parked bicycles generally requires some natural or formal surveillance, or else the provision of secure storage areas or compounds. Surveillance also reduces the risk of theft of cycle components.

Design principles

Street furniture

Provide cycle stands or similar street furniture in places with good natural surveillance.

This is appropriate in areas which attract members of the public on bicycles, such as shopping centres or public buildings like libraries, colleges, etc. To maximise natural surveillance, stands should be located in well-used places where there are many passers-by. Natural surveillance is also enhanced if the occupants in adjoining properties have a good view of the cycles.

It is not satisfactory to provide cycle stands in quiet, unsupervised places, such as side alleyways, at the back of buildings or in public car parks.

The design of cycle stands is discussed in Hudson (1982). Note, however, that single-function bicycle stands are not the only design solution. Railings, planting enclosures, benches, bollards, etc, are traditional features in the urban scene which can function well for securing a limited number of bicycles; but purpose-designed facilities are more appropriate where there are many cycles to be accommodated.

Secure cycle parks

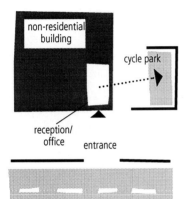

Provide cycle parks which are overlooked by offices or reception points or guarded by security staff.

The provision of secure cycle parks is an option for bicycles belonging to building users rather than the general public, for example, employees at factories, offices, etc, or students at schools and colleges.

These cycle parks should allow for bicycles to be locked to stands or racks.

The cycle parks should be located to discourage casual access by the general public.

Where to find out more

Research sources

McClintock, H (1992)
The Bicycle and City Traffic

Wheeler, A H (1989)
Cycle Theft Update

Good practice

Hudson, M (1982)
Bicycle Planning: policy and practice

■

13 LOADING AND UNLOADING

The areas around loading and unloading points should be designed to reduce the risk of theft, pilferage or even attacks.

The problem

Many security arrangements are designed to protect valuable goods when they are safely inside buildings but these measures do not help when goods are in transit. This is an issue in all buildings to a greater or lesser extent, but for retail and distribution buildings it is absolutely central. In these buildings, and in other large buildings where substantial quantities of goods are moved in or out on a regular basis, as much attention should be paid to security in transit as to establishing secure storage areas.

Where the risk occurs

The threat to goods when they are being loaded or unloaded in special loading areas comes mainly from employees. High street or commercial deliveries in public places are in theory at risk from anyone who is passing or waiting; but in practice goods are regarded as the responsibility of the carrier until they are safely within the recipient's building, so the risk is not borne by the building owner. If deliveries are expected on a large scale a special loading area would be provided, certainly in new buildings.

The special case of cash deliveries or collections from banks, and similar high security transport, is beyond the scope of this guidance.

Approaches to prevention

There has not been any published research on the specific security problems of loading and deliveries and so the following recommendations are based on a review of literature on current good practice.

The overriding strategy is to separate goods in transit, and the areas where they are handled, from other activities and spaces.

Design principles

Separate loading areas

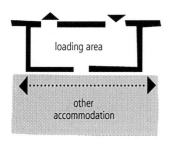

Provide loading doors and loading bays which are solely used for goods entering and leaving.

There should be no possibility of employees, much less the public, having unauthorised access to the loading areas.

Provide internal areas where goods can be received/dispatched and checked, which are only accessible to authorised staff. No routes between other parts of the building should pass through this area.

A desirable principle is to have separate doors and loading areas for goods coming in and goods going out. This reduces opportunities for diverting goods.

The provision of toilet facilities for incoming drivers reduces the need for visitors to enter the receiver's premises.

It is important that loading access should not double up as a staff entrance, nor even be used informally as such, in order to minimise staff pilferage.

Rubbish should not be taken out through loading areas.

Separate vehicle access

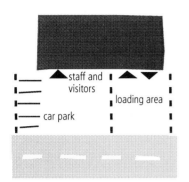

staff and visitors

loading area

car park

Provide a vehicle entrance and manoeuvering area which are dedicated to loading and unloading.

Separate loading areas may only be possible on large sites. The preferred site layout provides an enclosed external area outside the loading doors or loading bays. It should only be accessible to vehicles and people who are directly involved.

Car parking areas should be completely separated from delivery areas to make it harder for stolen goods to be concealed and removed from the site.

If an external loading area is fenced, gated and even controlled by a security officer it helps to prevent theft or pilferage during working hours and burglary outside working hours.

For smaller buildings where it is not feasible to have fenced delivery areas, it is recommended that loading bays should not open onto busy streets or narrow alleys.

Avoid hiding places

Minimise the opportunities for theft within loading areas.

Within the internal loading areas avoid nooks and crannies where goods could be concealed. The layout should maximise opportunities for surveillance.

It is desirable to provide distinct, secure areas for valuable, dangerous, damaged or returned goods. Good management also plays an important role here – if loading areas are disorderly it is harder to identify goods which have been diverted from their proper destination, either inwards or outwards.

A similar requirement occurs in §7 'Employee pilferage'.

Loading doors

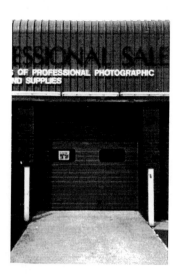

The design of external loading doors should be matched to the expected level of risk.

Loading areas are usually at the rear of buildings with relatively little natural surveillance, and are often accessible to vehicles outside working hours, so they are particularly vulnerable to sustained attacks by burglars.

In order that delivery doors can be kept shut except when loading or unloading is actually taking place, they should be easy to open and close, and provision should be made for approaching vehicles to be seen by staff inside the building.

Consideration should also be given to the issue of ramraiding (§3).

Where to find out more

Good practice

British Standard 8220: Part 3: 1990
Security of Buildings against Crime: Warehouses and Distribution Units

Fire Protection Association (1990)
Security Precautions: Site layout

■

14 FLY-PARKING

If a private car park or even a landscaped area is located where there is pressure on parking space, there is a need to exclude unwanted cars.

The problem

Fly-parking is not a criminal activity but it clearly becomes a nuisance because it either competes with the legitimate parking needs of an organisation or damages landscaping. While it may not be a problem immediately, it would be prudent to make provision at the outset of a design project so that the problem can be managed in future.

Where the risk occurs

The risk is likely wherever private on-site parking is provided in areas with severely restricted street parking. Typical locations include city centres and business districts, but other locations, for example close to commuter railway stations and sports stadia, may have particular problems. Also at risk is any land that may be attractive to people such as 'new age' travellers.

Approaches to prevention

There are a number of approaches which might be considered. Much depends on how actively the site access will be managed. For example, prestige hotels and business headquarters can rely almost entirely on the active deployment of doormen and security staff to control access to parking. Other businesses rely less on management and more on design and layout.

Design principles

There are several options depending on the likely pressure for fly-parking.

Define parking areas

Make clear separation between private parking and public spaces.

Avoid locating private parking spaces in positions where casual access is easy and unobtrusive. It is preferable to restrict access to clearly demarcated portals, gateways or driveways. Restrict pedestrian access in the same way, so that fly-parkers have to walk past positions where they may be challenged.

One-way barriers

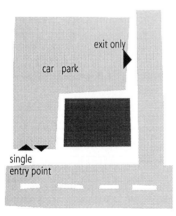

Provide one-way exit barriers which prevent cars entering.

If a car park has more than one access point, designate the one that is most effectively supervised as the single entrance; in order to prevent the other access points being used as entrances, install exit-only devices – for example, hinged ramps. This is simpler than full access control (and less effective), but makes it difficult for fly-parkers to slip in unobserved.

CIRIA SPECIAL PUBLICATION 115 · DESIGN FOR INHERENT SECURITY · 1995

Entry and exit control

Make provision for the installation of entry control barriers.

Access control systems with barriers are the only certain way of preventing fly-parking, and any private car park thought to be at risk should be designed so that barriers can be installed, either immediately or at a later date should they become essential. Barriers may be operated by pass cards or remotely by door/reception staff.

If access controls are used, allow a small supervised space for visitors to park temporarily to make enquiries at a reception point about parking arrangements.

Vehicle obstructions

Prevent access over, and parking on, landscaped areas.

Landscaped areas (soft or paved) next to streets/roadways in locations where there is pressure on parking, should be protected with low walls, hedges, ditches, timber posts, etc, to prevent vehicle access and parking.

Where to find out more

There does not seem to be any published research or design guidance on this security issue.

■

15 CUSTOMER ONLY ACCESS

There are many situations where businesses or public facilities need to welcome customers and visitors but discourage or prevent unacceptable use and activity.

The problem

Modern commercial business and public facilities often aim to attract customers and visitors, but their very openness can attract undesirable use and activity. Undesirable users may be merely wishing to shelter from the cold or rain. They may be rowdy youths or vagrants or children playing on skateboards, or prostitutes or even drug pushers. They may be people just taking a short-cut. Essentially, they are seen by building managers as damaging to their proper business or function.

Where the risk occurs

All publicly accessible buildings can suffer from undesirable users, but the risk is mostly determined by the particular location and surroundings of the building in question. Further information on such risks might be obtained from neighbouring owners and the police.

Typical locations include bus and railway stations, shopping malls, public car parks, city centre hotel lobbies, hospitals, multi-tenanted office buildings, libraries, art galleries, etc, as well as semi-public space in front of buildings.

Approaches to prevention

Most problems of non-customer access in publicly accessible areas of buildings can in principle be controlled by security personnel, doorkeepers or other staff, but this is not feasible in many situations.

There seems to be a number of ways in which the design of the physical environment can reduce the problem. The first is to exploit informal supervision. Where it is possible to design public areas to include facilities such as enquiry points, shops and cafés, the staff from these can be encouraged to act in an informal policing role.

Another approach is to minimise attractiveness to non-customers. Avoid creating unnecessary attractive features, for example by eliminating seating areas and large unused space or alcoves which might attract loitering people or vagrants. This approach might be used in a railway station but not would not be appropriate in a comfortable hotel lobby/lounge area.

In some situations the above strategies are insufficient. For example, it might be impractical or too costly to give adequate supervision, or the pressure for access by un-desirable users may be too great. Here it might be better to minimise the area open to free public access.

'This boy,' says the constable, 'although he's repeatedly told to, won't move on.'
'I'm always a-moving on, sir,' cries the boy, wiping away his grimy tears with his arm. 'Where can I move to!'
'Don't come none of that. My instructions are, that you are to move on. I have told you so five hundred times.'
'But where?' cries the boy.

Dickens *Bleak House* (1853)

Design Principles

accommodation

reception

cafe

shop

entrance

street

Supervision in public areas

Provide surveillance by the strategic location of staff at reception desks, information points, shops, bars, coffee shops, etc, and ensure that people entering the building have to pass through supervised areas.

Surveillance can be achieved by creating areas which are under the specific care and control of particular staff, such as a café or shop or a temporary sales pitch for a new car or a double-glazing company.

Ensure that access to stairs or lifts is only possible for the public by first moving through well supervised areas, such as past a reception desk or security point.

Provide access to public toilets, telephones and similar conveniences only through well supervised areas, such as a restaurant or on upper floors only.

Note that some situations may require more active patrolling by police or security staff.

Discourage non-legitimate use

Avoid design features that would attract non-legitimate users.

Avoid, for example, providing comfortable seating. It is particularly important to avoid these features in areas which are poorly supervised. Note efforts to design anti-vagrant seating such as rounded stone benches with no backs or individual seats.

An example in the police guidance *Secured Car Parks* (Association of Chief Police Officers, 1992b) suggests making the surface of vehicle ramps too rough for skateboarding.

Avoid or eliminate short cuts through public areas of buildings.

Limit area of public access

Where 'softer' methods of control are felt to be inadequate, provide means of physical access control.

Physical access control can be achieved in many ways, for example by turnstiles operated by pass cards in office buildings, or by coins to gain access to public toilets.

Public areas can sometimes be locked outside business/operating hours.

In transport interchanges it may be possible to locate seating and other facilities inside ticket control barriers, to restrict their use to legitimate travellers.

Where to find out more

Good practice

Association of Chief Police Officers (1992b)
Secured Car Parks
Includes some advice on discouraging the use of car parks as play areas.

■

16 STAFF ONLY ACCESS

A common problem is managing access to a building or parts of a building in order to provide access for staff while excluding the general public, customers or unwanted intruders.

The problem

The need to exclude the general public, customers or unwanted intruders is usually to protect the building, its occupants and/or its contents from a variety of potential crime and nuisance behaviour.

Examples of buildings where this issue is important include offices: free movement and access by staff is clearly necessary, but uncontrolled access by others 'opens the door' to theft, nuisance, loss of confidentiality and even commercial espionage or terrorism.

Another example is in hospitals where patients and staff must be protected, particularly at night. There is a need for medical staff to be able to move about the hospital quickly and easily, but without the risk of an ill-intentioned intruder gaining access to attack staff or abduct a child, etc.

Where the risk occurs

The risk is probably more general than is normally realised. Even the most strongly customer-oriented environments, such as retail and service facilities, always have some areas that are reserved for staff only.

Some activities require much closer security than others. Staff access to most offices and industrial premises requires moderate control, but where high-value goods or materials are handled or where sensitive or confidential document-ation is held, greater security is needed. Access to offices dealing with financial and government matters typically require comprehensive access control.

The problem can be particularly difficult in multi-tenanted buildings.

Approaches to prevention

There are two main approaches to the exclusion of intruders. The first is to employ some form of manned security control at all possible entrances. Staff will either be recognised or required to show some formal means of identification before they are allowed to enter.

The main advantage of supervised access is that the degree of security control can be varied to suit particular conditions, and legitimate visitors can also be allowed in. Some situations need no more than a receptionist to keep a watchful eye on those coming and going; elsewhere access can be tightly controlled by specialist security staff, who may even carry out searches of bags, briefcases, etc.

The second approach is to provide locked access points, accessible to staff with a key, swipe card or access number codes. Such entrances need to be designed with care to reduce the risk of attacks on staff entering or leaving.

Further security may be gained by controlling the locking mechanism from within the building. Here there is a need to provide surveillance of the entrance area to check that it is safe to release the lock. Surveillance of the entrances can be directly through glazing or remotely with the use of a CCTV camera.

The House should not have above one Entrance, to the Intent that nobody may come in, nor any thing be carried out, without the Knowledge of the Porter.

Alberti *The Ten Books of Architecture* (1472)

Design principles

Supervised entrances

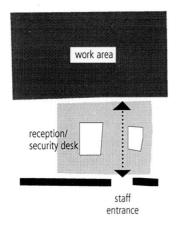

Where entrances are permanently staffed, all access routes into the building should pass immediately in front of a reception/ security desk.

It may sometimes be necessary to restrict the flow of access by temporary or permanent barriers or rails, so that individuals can be checked more methodically. The use of low-level turnstiles may be helpful.

Care needs to be taken to ensure that other fire exits, loading doors, etc, cannot be used to gain access from outside. Self-closing and self-locking mechanisms may be useful. Consideration might be given to the use of warning signals to indicate when such entry/exit points are opened.

Provision for visitors

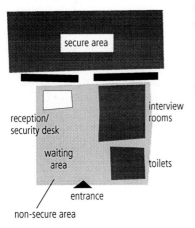

Where visitors have to wait before moving into a secure area, provide for their needs outside the secure area.

Normally, this involves locating the reception point and waiting areas with seating, and any other facilities for visitors, on the entrance side of the reception/security point. This area should be under the supervision of reception or security staff.

An extension of this principle is the provision of public access facilities, such as interview and conference rooms, in the non-secure zone.

Staff operated access

Staff-only access doors should be fitted with locking systems that can only be operated by approved personnel.

There is a wide range of locking systems from conventional key-operated locks to others operated by electronically coded cards or by keypads.

There is no reason why locked entrances should compromise fire escape requirements, as they need only be secured against free access from the outside.

Consideration should be given to the surveillance of these entrances by staff from within the building. Ideally these entrances should be overlooked by windows. Alternatively, a CCTV camera may be useful.

There should be sufficient space around an entrance for staff to feel confident that they are not about to be attacked or crowded as they open the door.

Where to find out more

This is a security issue which appears to have received little attention from research or published design guidance.

■

17 FLY-TIPPING

Areas of open land, whether landscaped or unused, can be illicitly used for tipping waste, resulting in loss of amenity, nuisance and sometimes pollution.

The problem

The ever-increasing need for the disposal of waste creates constant pressure for dumping. Despite the development of official waste depots and the availability of well-managed skip hire services, there is always a minority of people who disregard social conventions and dump waste on any available open ground.

Materials dumped include debris from minor building work, old cars and unwanted household appliances or furniture. Not only is the dumped material a disamenity in itself, but it can also contribute to a more general litter problem through lighter material being blown about. Waste may also be spread about by children.

Where the risk occurs

Dumping can occur wherever there is open or unused land. The risk may be greater in urban areas but also afflicts country areas. Dumping tends to occur close to vehicle access points which have little surveillance from main roads or surrounding buildings.

Approaches to prevention

It is generally believed that well-kept grounds are less likely to attract fly-tipping. In these situations some light dumping may not be a serious problem because it will be cleared in the normal course of street cleaning or maintenance work by land owners.

The more serious problem comes when large quantities of waste are dumped from vehicles, including trucks and vans. The approach most likely to prevent such dumping is to close off access by vehicles.

Unused land might be closed off by boundary fencing, hedges, walls, etc, but where areas are landscaped for amenity the use of fencing and walls would be unacceptable. Here, there are alternative landscaping devices which can be used to prevent vehicle access.

Design principles

Fencing to yards and unused land

Unsupervised areas of open land near to buildings should be protected by fencing, dense hedges or walls.

This applies to areas of open or unused land which do not have amenity value.

Boundaries should not be set too far back from roadways as this also leaves space for dumping. Gates need to be in good order and securely locked.

Vehicle obstruction by landscaping

Open landscaped areas should be protected with devices to prevent vehicle access.

Examples include the use of stout timber posts set a few feet apart along the side of roads, ditches, banking or heavy planting. Such devices are also used to prevent parking in landscaped areas or woodland and do not interfere with access on foot.

Where to find out more

There does not appear to be any research or design guidance on this aspect of inherent security.

■

18 LOITERING GROUPS OF YOUTHS

Groups of youths loitering in public areas can be a source of fear for some members of the public and are often considered to be undesirable for business reasons.

The problem

A survey of a Midlands shopping centre showed that gatherings of youths were an even greater source of concern to traders than the level of shoplifting: 'It was found that incidents of nuisance, not crime, were the predominant problem and that these incidents stemmed from a conflict of interest between shoppers, retailers and the young people who were using the centre as a meeting place' (Phillips and Cochrane, 1988).

Conflict is the key word. Few people would question that young peoples' casual gathering in groups is a natural and normal pattern of behaviour, but such gatherings can be threatening; and it has been found in surveys that they contribute significantly to fear of crime (LaGrange et al, 1992).

Generally the nuisance caused by loitering is not a crime at all, but it is linked to crime through being a violation of normally accepted social behaviour. These kinds of violations of accepted standards are sometimes called *incivilities*, as opposed to crimes.

Among corner gangs of aggressive, alienated urban youth the notion of 'kicks' has its fullest bearing. Here the community itself is transformed into a field for action, with special use made of peers, unprotected adults, and persons perceived as symbols of police authority.

E Goffman *Interaction Ritual* (1967)

Where the risk occurs

Groups of youths gather in places with a feeling of liveliness (whereas vagrants or travelling people often seek obscure, sheltered corners), reflecting their desire for a stimulating environment which they can occupy at no cost. There is no reason to assume any intention to commit crimes or intimidate law-abiding citizens.

Groups of youths are drawn to certain physical features in the environment. A change in level up to about 900mm high is a natural seat. A barrier, rail, parapet or wall up to about 1350mm high is a place to lean. A wall or an internal corner can be leant on or used for some minimal shelter. These physical features can be considered as 'magnets' for loitering. Where they are present in lively, public areas they will probably attract groups of youths.

Wide column bases in a suburban shopping parade attracted teenage gangs, causing a nuisance to residents of the flats above the shops. The column bases were removed in refurbishment works. However, no alternative magnets were provided and the youths now sit on the entrance doorsteps to the flats.

CIRIA SPECIAL PUBLICATION 115 · DESIGN FOR INHERENT SECURITY · 1995

Approaches to prevention

The most direct response is for police, security staff or others in authority to request groups to 'move on'. However, loitering is a problem with a high probability of displacement – that is, if a group is prevented from loitering in one place it is quite likely to move to somewhere else, rather than dispersing.

Physical deterrents are sometimes used to prevent magnets being used as a base for loitering, but they seem to demonstrate a failure of inherent design. In general, deterrent design is not the way to go: it implies that a magnet has been created which has to be neutralised. Far better to avoid magnets, or locate them where they will not lead to conflicts.

Design principles

Location of 'magnets'

Avoid placing 'magnets' in sensitive locations.

The objective is to prevent conflicts between groups of youths and other members of the public, so avoid placing design or landscape features that could act as magnets for groups of youths in locations where conflicts would be likely.

A positive approach would be to ensure that design or landscape features that would act as magnets are located in places where there would not be conflict with other users, rather than attempting to eliminate all such design features completely. A location that was too remote from any action would, however, fail as an effective magnet.

Gallery design

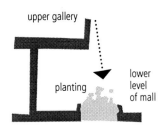

upper gallery

planting

lower level of mall

Design galleries to discourage nuisance behaviour.

'Gallery abuse' is a pattern of nuisance behaviour which occurs in multi-level shopping centres. Youths on an upper level lean over balustrades which are directly above public circulation areas and drop objects or deliberately antagonise members of the public on the lower level (Poole, 1991).

It can be avoided by ensuring that balustrades or parapets on galleries are located immediately above planting or other inaccessible areas which divert the public on the lower level. Alternatively, by placing planters or other obstructions against the upper level balustrade, youths will not be able to lean over.

Where to find out more

Research sources

LaGrange, R L, Ferrado, K F and Supancis, M (1992)
'Perceived risk and fear of crime: social and physical incivilities'

Phillips, S and Cochrane, R (1988)
Crime and Nuisance in the Shopping Centre

Good practice

Poole, R (with Donovan, K) (1991)
Safer Shopping

19 WILFUL DAMAGE

Damage to buildings can be caused by accident and wear and tear, but sometimes it is the result of a wilful act which might be prevented.

The problem

All buildings have to be designed to withstand normal wear and tear and accidental damage. Normal wear and tear often involves heavy use and careless behaviour. Design may also fail to provide adequate facilities which in turn leads to misuse of buildings or parts of buildings.

Vandalism is often confused with these normal design and maintenance problems, but it has one important difference. It is essentially malicious or wilful behaviour and may therefore be preventable.

Evidence from crime surveys frequently shows that this is one of the less well reported crimes. Vandalism in non-residential settings is not identified in national crime statistics nor is it part of the British Crime Survey. In surveys of school arson (Burrows et al, 1993) and crime on industrial estates (Johnston et al, 1990) incidents of vandalism appear to approach the number for burglaries.

Research shows that most vandalism is caused by youths and children (see Clarke, 1978). The behaviour is play related and perhaps part of the process of maturing. Where youths and children gather, play or hang out, their attention is drawn naturally to objects in the environment. If these objects are weak or known to be weak, they will be tested or experimented with. If unsupervised they may soon be tested to destruction. Perhaps, unlike much crime, the behaviour is not planned but grows out of the situation – out of exploring the environment.

In rough neighbourhoods the lower part of the door should be covered with strong sheet iron, able to bear successfully the perpetual kicking which it is sure to receive.

E R Robson *School Architecture* (1874)

Damage is usually limited to what can be done by kicking things, throwing stones or forcing things. In some cases this might include dropping stones, etc, from a height. The behaviour is 'natural', particularly for young males. A pebble thrown on a beach is a harmless and normal activity. Even on waste ground this may be harmless but in the urban environment it can lead to broken windows, street lamps, etc.

Where the risk occurs

Isolated sports pavilions are particularly at risk of vandalism – they are in public areas well known to youths and children, and often have minimal supervision when not in use.

Although the problem is common and extensive, the risk is by no means found everywhere. It is characteristic of places that are easily accessed by youths and children but less so by adults. This is why it was such a major problem for much public sector housing in the 1960s and 70s. Housing estates were designed with vast areas of open space, garage courts, entrance lobbies, lifts, staircases, etc, which attracted a good deal of damage.

The problem tends to be limited to similar situations in most commercial, retail and industrial buildings with back yard areas and unsupervised communal access areas. The problem is more serious in buildings where youths and children congregate, such as schools, transport and recreational facilities, and perhaps some car parks.

Sometimes features such as outdoor sculpture and fountains will attract special attention.

Approaches to prevention

Much emphasis for prevention is placed on the need for supervision and control of children and young people. Many believe that the solutions to vandalism depend on the good management of buildings. If management means that the buildings and their surroundings are carefully supervised or guarded, then there is little doubt that the vandals can be kept at bay. This is certainly part of the role of people like factory gatekeepers and hotel doormen.

CIRIA SPECIAL PUBLICATION 115 · DESIGN FOR INHERENT SECURITY · 1995

However, close guarding of premises is not always practical. Design can play an important part. What is suggested is that designers attempt to protect the outside and inside of buildings by adopting the following strategy.

- Wherever possible, eliminate areas in which youths and children are likely to have access and play or hang out.

- If such areas cannot be avoided, provide as much natural surveillance as possible.

- If this is not possible, harden the environment beyond what would be needed to deal with expected wear and tear.

Always repair damage as soon as possible.

Design principles

Provide external surveillance

Wherever possible provide windows to overlook areas surrounding the building.

Effective supervision from surrounding buildings is only achieved during the hours when the buildings are occupied.

The windows must be designed to give a view out – not, for example, high level windows.

Avoid access to unsupervised areas

Unsupervised areas around buildings which might be attractive to youths or children should be made inaccessible.

Areas which suffer from vandalism usually lack adult presence or supervision. Keeping youths and children out of these areas may require yards, loading areas, garden courts, etc, to be fenced or walled off.

An alternative approach in some situations may be to provide rough surfaces, such as cobbles, which are uncomfortable to walk on and deter ball games, cycling, etc.

Protect immature planting

Fence off areas of immature landscaping.

Young trees and other planting and landscaping should be protected by fencing until mature.

Harden unsupervised internal areas

Unsupervised internal rooms used by youths and children should be designed with robust furniture, fittings and equipment.

Robust design is appropriate in, for example, lavatories, lobbies, stairwells, lifts, etc. Many manufacturers have designed equipment which is intended to be vandal-resistant. This topic is discussed in Cheetham (1994).

Pedestrian through routes

Property along routes likely to be used by youths and children should be protected by space or physical barriers.

Buildings beside pedestrian routes through non-residential areas which are likely to be used by youths and children, for example routes to public transport, should be protected from possible damage. This might be done with fencing, walls or railings, etc, or with mature landscaping and particularly hedging or other planting.

Where to find out more

Research sources

Burrows, J, Shapland, J, Wiles, P and Leitner, M (1993)
Arson in Schools

Clarke, R V (ed) (1978)
Tackling Vandalism

Johnston, V, Leitner, M, Shapland, J and Wiles, P (1990)
Crime and Other Problems on Industrial Estates

Good practice

Cheetham, D W (1994)
Dealing with Vandalism: a guide to the control of vandalism

Department of Education and Science (1987)
Crime Prevention in Schools: practical guidance

Sykes, J (ed) (1979)
Designing against Vandalism

Tyne and Wear County Council Architects Department (undated)
Reducing Vandalism by Design and Management

Ward, C (ed) (1973)
Vandalism

Zeisel, J (1976)
Stopping School Property Damage: design and administrative guidelines to reduce school vandalism

■

20 GRAFFITI

Some people find graffiti 'art' beautiful but it is a nuisance to most building owners. Worse still, it can signal decline and neglect and thereby invite other crime or nuisance behaviour.

The problem

Graffiti can display a kind of artistic creativity, but unfortunately coupled with disrespect for others' property.

Unlike most crime or delinquent behaviour, everyone will have seen large graffiti, whether painted onto motorway bridges, along urban railway tracks or on the walls of pedestrian subways.

Whilst there is a considerable literature on graffiti and many proprietary products for removing graffiti from walls (see, for example, Wallace and Whitehead, 1989), there is less clear guidance on how to avoid graffiti, or indeed, on how far it can be prevented by good design.

However carefully surface finishes are selected to discourage painting, spraying, etc, they are still capable of being marked. Rough surfaces may deter graffiti but if sprayed they are almost impossible to clean. Smooth, cleanable and specially sealed surfaces still attract renewed attacks. The question here is how to avoid graffiti in the first place.

Where the risk occurs

I wonder, O wall, that you have not collapsed under the weight of all the idiocies with which these imbeciles cover you.

Graffiti found at Pompeii

Graffiti can appear on any flat vertical surface which is accessible to youths and children, and which is little supervised.

Outside locations include almost any wall that can be seen prominently from the street or from public transport, including rail and motorways. Walls around open spaces

Walls alongside railway lines are a common target – typically such areas lack surveillance but offer visibility for the completed graffiti.

in which youths and children hang out or play are also favourite locations. Another type of surface which attracts graffiti is the security shutters to shops.

Interior locations tend to have smaller graffiti, which appear in unsupervised areas, such as lavatory walls, lift cars and stair lobbies.

Approaches to prevention

The importance of management of building facilities to avoid graffiti is argued by many. For example, the elimination of the notorious graffiti on trains in the New York Subway was largely achieved by a determined management policy. Trains were not allowed into service unless they were clean. Eventually the graffiti 'artists' gave up, because their handiwork was never seen (Sloan-Howitt and Kelling, 1992).

However, there is good reason to believe that if the opportunity is removed in the first place, graffiti will not be a problem. Graffiti can only be created if there are reachable surfaces to write on or paint. It does not usually appear in areas which clearly belong to people or organisations, for example, where there is some surveillance and where there the owners or legitimate users of the buildings are present (compare with the defensible space theory of Oscar Newman, 1973).

THE WAGES OF SIN IS DEATH - BUT THE HOURS ARE GOOD.

Graffiti in Sheffield, from N Rees *Graffiti 3* (1981)

A simple example illustrates the idea of removing the opportunity for graffiti. The stations of the metro system in Washington DC were designed carefully with the prevention of crime and vandalism in mind. The walls and roof of the stations are formed from arched concrete coffer units. This design provides no suitable surface for graffiti, but, just in case, the platforms are separated from the walls by a deep servicing channel which prevents anyone reaching the main walls of the stations. The Washington metro is virtually free of graffiti.

Design principles

The following design principles are offered as a means of removing or reducing the opportunity for graffiti to appear in and around buildings.

Windows overlooking external grounds

Wherever possible provide potential surveillance of areas around buildings.

Graffiti does not generally occur on surfaces where there is the potential for surveillance from surrounding buildings.

Avoid plain flat boundary walls

Where boundary walls or fences face the street or publicly accessible open space, use alternative structures to flat vertical surfaces.

Examples of such alternatives would include wire mesh fencing and railings.

Some heavily sculptural surfaces may discourage graffiti, but may provide an opportunity for other forms of abuse in a more high risk setting.

Even if flat surfaces are used, it might be possible to protect them by close planting of hedges or climbing plants. Graffiti are not usually sprayed onto foliage.

Alternative boundaries can be formed with landscaping, such as hedging or planted mounds.

Use easy clean surfaces internally

Where there is a risk of graffiti within a building, use easily cleaned surfaces.

Surfaces designed for easy cleaning should be used in places such as lifts, stairwells, lobbies and lavatories, wherever there is poor supervision and the possible presence of unsupervised youths or children. More detailed guidance may be found in Wallace and Whitehead (1989).

Good maintenance and cleaning are always necessaøry in these areas.

Internal surveillance

Where possible, introduce glazing to improve internal surveillance of potential trouble spots.

Privacy will constrain some use of internal glazing, but there are examples of relatively recent design trends which enhance surveillance, such as the use of glass lifts.

Where to find out more

Research sources

Clarke, R V (ed) (1978)
Tackling Vandalism

Newman, O (1973)
Defensible Space: people and design in the violent city

Good practice

Cheetham, D W (1994)
Dealing with Vandalism: a guide to the control of vandalism

Department of Education and Science (1987)
Crime Prevention in Schools: practical guidance

Sloan-Howitt, M and Kelling, G L (1992)
'Subway graffiti in New York City: "Gettin Up" vs "Meanin It and Cleanin It"'

Sykes, J (ed) (1979)
Designing against Vandalism

Tyne and Wear County Council Architects Department
(undated)
Reducing Vandalism by Design and Management

Wallace, J and Whitehead, C (1989)
Graffiti Removal and Control

Ward, C (ed) (1973)
Vandalism

Zeisel, J (1976)
*Stopping School Property Damage: design and administrative
guidelines to reduce school vandalism*

■

21 ARSON

Much fire-setting by youths and teenagers is similar to vandalism and the risk can be controlled to some extent through design.

The problem

Arson is a concern of the insurance industry as it appears to be increasingly the cause of fires. Furthermore, fires started deliberately tend to be more serious as they are often started with accelerants and in more than one place.

Although the motive for arson can be fraud, sabotage by disaffected employees and others, or a racial attack, it is generally believed that this represents only a small proportion of the problem. More typically it appears to be associated with vandalism and burglary; the offenders being young teenage males or even younger children (CFPA Europe, 1989).

Where the risk occurs

The building types which seem most at risk from arson are schools and retail premises, but by comparison with burglary the risk is small. A recent survey of arson in schools found that in 1990 one in eight schools (12%) experienced at least one arson attack; and data for 1992 showed that 19% of schools had experienced arson in a 14-month period (Burrows et al, 1993). However, the 1992 data showed that 70% of schools had experienced burglary (see table on page 142).

The survey found that larger secondary schools were more at risk than smaller primary schools. Schools affected by other crimes such as burglary, theft and vandalism were more likely to suffer from arson as well. Targets of arson tended to suffer repeated attacks.

Turning to retail premises, a recent report calculated the risk of arson as one attack per 100 retail premises per year (Burrows and Speed, 1994). The figure rose to three per 100 per year for grocery retailers, which may partly reflect some of the problems of racial attacks associated with small shops. Again, the level of risk for arson is very small compared with crimes such as burglary and criminal damage (57 and 54 attacks per 100 retail premises per year respectively).

Commercial and industrial premises are also considered to be a common target for arson attacks. However, a survey of crime on industrial estates showed the risk to be small (Johnston et al, 1990). The survey covered 584 premises and identified 562 criminal incidents in a two-year period – only one was an arson attack. The figure for burglary was over 100.

An important factor in determining the risk of arson is likely to be the general character of the surroundings. For example, if a site borders a run-down housing estate or derelict land used by teenagers and children the risk will be greatly increased.

Approaches to prevention

The problem of preventing damage from arson is dealt with in current guidance literature in, broadly, two ways. The first approach is to reduce damage once a fire has been started, and the second approach is to prevent people setting fire to buildings in the first place.

In general, the control of fire spread is the same for all fires and is not dealt with here in detail. However, measures can be summarised as follows:

- fire resistant materials in the building fabric

- compartmentation of the building

- automatic fire detection and alarm systems

- automatic fire extinguishing systems, such as sprinklers.

The second approach is to prevent arson. There are many possible motives for malicious fire-setting. Attacks motivated by fraud, sabotage and the covering up of other crimes, as well as arson by employees, seem to be beyond the influence of building design. If these forms of arson are to be controlled, they require policing by the insurance industry and the careful management of staff by owners of businesses.

The greater part of the problem of arson is seen to be associated with the behaviour of teenagers and younger children. In particular, the schools survey referred to above showed that headteachers believed that 50% of school arson was simply acts of vandalism and a further 30% was thought to be the result of 'play behaviour'. The remaining incidents were attributed to disturbed behaviour and grudges against the school.

A further point to emerge about school arson is that most fires are started on the outside of the building. This suggests that most arson attacks could be prevented if teenage males and younger children could be kept away from school buildings outside school hours. Headteachers believed that perimeter fencing was the most important physical improvement that could be made to reduce the risk of arson.

In the absence of further research, it seems reasonable to assume that access control to grounds around more isolated buildings (such as schools) would reduce the risk of arson. It would be helped by surveillance from staff, such as caretakers or security personnel, and from neighbouring streets.

Some existing guidance goes further and adopts virtually all the common security measures associated with burglary prevention; see for example the *Arson Dossier* published as a European initiative (CFPA Europe, 1989). These measures might seem excessive merely for arson prevention, but their use would be justified on the basis of both burglary and arson prevention jointly.

Two more specific areas of risk which can be countered by preventive design are sometimes referred to in guidance literature. They are attacks on entrance doors of small businesses, and fires started in unprotected rubbish and waste.

Design principles

Discourage access by teenagers and children

Any unsupervised area adjacent to vulnerable buildings should be protected by a fence, wall or hedge robust enough to discourage illicit access by teenagers and children.

The main target for such treatment would be schools, particularly secondary schools, but other campus-like developments might benefit from this principle, including industrial estates and hospitals.

The robustness of the boundary is not so much to make access impossible but to discourage routine access across open areas or through neglected breaks in the fences, etc. Where fencing is adjacent to roads and other open ground it will aid surveillance to make it transparent above one metre using vertical railings or mesh.

Gateways need to be closed and locked when the premises are not in use. There is probably no need for heavy fortification.

Lighting and surveillance equipment may be of use providing there is some human presence such as a caretaker or security staff to act as guardians.

Waste storage and refuse containers

Areas used to store waste, particularly waste paper and other flammable materials, should be contained within a yard or store.

An enclosed yard with a wall or fence at least 2m high would probably be sufficient to keep out youths and children. Such a yard should be locked when the building is not in use. It may be an advantage to keep these storage areas away from the main buildings because they are vulnerable to fire raising.

CIRIA SPECIAL PUBLICATION 115 · DESIGN FOR INHERENT SECURITY · 1995

Letter boxes

Letter boxes and associated areas at entrances to commercial and industrial buildings should be fire resistant.	Some mention is made in the literature of incendiary devices used to attack doors, such as small businesses like shops or workshops. In such buildings consideration might be given to the design of entrance areas, including letter boxes or postal chutes, etc, to ensure that they are constructed only in fire resistant materials.

Where to find out more

Research sources

Burrows, J, Shapland, J, Wiles, P and Leitner, M (1993)
Arson in Schools

Burrows, J and Speed, M (1994)
Retail Crime Costs 1992/93

Johnston, V, Leitner, M, Shapland, J and Wiles, P (1990)
Crime and Other Problems on Industrial Estates

Good practice

Arson Prevention Bureau (1992)
Prevention and Control of Arson in Industrial and Commercial Premises

Arson Prevention Bureau (1993)
How to Combat Arson in Schools

CFPA Europe (1989)
Arson Dossier

Fire Protection Association (1992)
Prevention and Control of Arson in Warehouse and Storage Buildings

■

PART 3

SECURITY ISSUES IN SELECTED BUILDING TYPES

Most of the security issues described in Part 2 are relevant to many different building types. To assist those involved in particular projects, the following sections in Part 3 identify the issues which are most likely to be applicable in a number of important non-residential building types. As well as presenting a checklist of security issues, additional notes are added where necessary to draw attention to research findings and sources of information that are specific to the building types.

SHOPS AND STORES

TYPICAL SECURITY ISSUES

During normal trading hours

2	Smash and grab raids
4	Cash robbery
5	Violence to staff
6	Shoplifting
7	Employee pilferage
8	Customer belongings
9	Personal belongings at work
11	Parked cars (customer car parks)
13	Loading and unloading

Outside normal trading hours

1	Common burglary
2	Smash and grab raids
3	Ramraids

See the decriptions of each security issue in Part 2

Where to find out more

Research sources

See §6 'Shoplifting' (page 55).

Good practice

See §2 'Smash and grab raids (page 36), §3 'Ramraids' (page 40) and §6 'Shoplifting' (page 56).

■

MALLS AND SHOPPING CENTRES

TYPICAL SECURITY ISSUES

In addition to those which apply to shops and stores

10	Other theft from premises
15	Customer only access
18	Loitering groups of youths
19	Wilful damage

See the descriptions of each security issue in Part 2

Where to find out more

Research sources

Brantingham, P L, Brantingham, P J and Wong, P (1990)
'Malls and crime: a first look'

Phillips, S and Cochrane, R (1988)
Crime and Nuisance in the Shopping Centre: a case study in crime prevention

Good practice

Poole, R (with Donovan, K) (1991)
Safer Shopping
A comparative study of shopping malls in the UK and abroad.

■

OFFICE BUILDINGS

TYPICAL SECURITY ISSUES

1	Common burglary
4	Cash robbery
5	Violence to staff
7	Employee pilferage
9	Personal belongings at work
11	Parked cars
12	Bicycle parking
14	Fly-parking
16	Staff only access

See the descriptions of each security issue in Part 2

Commentary

A study of tenant satisfaction in newly built office buildings showed that security was the aspect of design that caused greatest dissatisfaction (Vail Williams, 1990). Unfortunately this survey did not reveal what particular security problems had been experienced. However, it clearly suggests the need for both further research and improved designs.

Where to find out more

Research sources

Vail Williams (1990)
The Occupiers' View: business space in the '90s

Good practice

Association of Chief Police Officers (1992a)
Secured by Design: Commercial
This booklet is available from local police crime prevention services (ask for the Architectural Liaison Officer).

British Council for Offices
Six datasheets on: office crime, bombs and terrorism, office car parking, CCTV, disaster recovery, business parks.

British Standard 8220: Part 2: 1987
Security of Buildings against Crime: Offices and Shops

■

INDUSTRIAL ESTATES

TYPICAL SECURITY ISSUES	
1	Common burglary
7	Employee pilferage
9	Personal belongings at work
11	Parked cars
12	Bicycle parking
13	Loading and unloading
14	Fly-parking
17	Fly-tipping
19	Wilful damage
21	Arson

See the decriptions of each security issue in Part 2

Commentary

A crime survey of 584 units on 43 industrial estates, mainly in the North of England, provides some indication of the levels of risk experienced by businesses (Johnston et al, 1990). The survey recorded 562 incidents of crime in a two-year period, and the data are summarised in the following table. The data are expressed as the number of incidents per 100 industrial units per year (see page 126).

Overall, the risk of crime other than burglary, theft and vandalism is small. Only one incident each of arson, assault and fraud was recalled in the two-year period.

Issue	Crime	Incidents per 100 units per year
1	Burglary	10
	Attempted break-in	7
5	Assault	<1
	Threats	1
7	Theft by employees	1
9/10	Theft	10
11	Theft of vehicles	2
	Theft from vehicles	2
	Damage to vehicles	3
19	Vandalism	8
21	Arson	<1
–	Fraud	<1

The authors of the study emphasise that the crime risk on estates can vary a good deal. The estate with most crime averaged about 500 crimes per 100 units per year, whereas the lowest crime estate averaged less than ten crimes per 100 units per year. It seems likely from the analysis in the report that the main reason for higher crime rates was the size of the estate and the units on it: larger estates with larger units employing more workers tended to have higher rates of crime. The authors comment that the more open estates with large amounts of empty space between the units had a greater risk of crime, particularly burglary.

Two other factors were found to be important. One was the general tidiness or scruffiness of the estate – the less well kept, the more crime. And estates which were located away from other development in open country were largely free of crime, whereas estates with the most problems were adjacent to both run-down public sector housing and to open land, no doubt providing both offenders and ease of access.

Where to find out more

Research sources

Johnston, V, Leitner, M, Shapland, J and Wiles, P (1990)
Crimes and Other Problems on Industrial Estates

Good practice

Arson Prevention Bureau (1992)
Prevention and Control of Arson in Industrial and Commercial Premises

British Standard 8220: Part 3: 1990
Security of Buildings against Crime: Warehouses and Distribution Units

Fire Protection Association (1992)
Prevention and Control of Arson in Warehouse and Storage Buildings

■

HOTELS

TYPICAL SECURITY ISSUES

Public areas

5	Violence to staff
10	Other thefts from premises
15	Customer only access

Guests' bedrooms

1	Common burglary
8	Customer belongings

Sports and leisure facilities

8	Customer belongings

Generally (including staff only areas)

1	Common burglary
4	Cash robbery
7	Employee pilferage

Car parking

11	Parked cars

See the descriptions of each security issue in Part 2

Where to find out more

Off the guest room there should be a repository, where the guest might hide his more personal or precious belongings, and retrieve them as he wishes.

Alberti *The Ten Books of Architecture* (1472)

There does not appear to be any published research on the specific security problems of hotels.

■

LEISURE/SPORTS BUILDINGS

TYPICAL SECURITY ISSUES	
4	Cash robbery
5	Violence to staff
8	Customer belongings
11	Parked cars
12	Bicycle parking
18	Loitering groups of youths
19	Wilful damage

See the descriptions of each security issue in Part 2

Commentary

The question of controlling crowd behaviour at football stadia has been discussed in Lord Justice Taylor's report following the disaster at Hillsborough (Taylor Report, 1991).

Much of this has been subsumed under other aspects of safety and design of stadia. A number of guidelines have been issued by the Football Stadia Advisory Design Council.

Where to find out more

Good practice

Football Stadia Advisory Design Council
Some FSADC guidelines are relevant to security.

Sports Council (1990)
Designing for Safety in Sports Halls: Security

Taylor Report (1991)
Hillsborough Stadium Disaster Final Report

■

PUBS, BARS AND RESTAURANTS

TYPICAL SECURITY ISSUES	
1	Common burglary
4	Cash robbery
5	Violence to staff
8	Customer belongings
10	Other thefts from premises
11	Parked cars

See the descriptions of each security issue in Part 2

Commentary

Keeping the Peace: a guide to the prevention of alcohol-related disorder (MCM Research, 1993) sets out some general principles for the design of public houses. The guidance is based on research and deals with general approaches to design and layout. It makes a number of suggestions about the design of licensed premises.

Where to find out more

Good practice
MCM Research (1993)
Keeping the Peace: a guide to the prevention of alcohol-related disorder

■

TRANSPORT INTERCHANGES

TYPICAL SECURITY ISSUES	
4	Cash robbery
5	Violence to staff
8	Customer belongings
11	Parked cars
12	Bicycle parking
15	Customer only access
18	Loitering groups of youths
19	Wilful damage

See the descriptions of each security issue in Part 2

Commentary

Some research into violence to staff (§5) has been carried out on the London Underground in relation to the design of ticketing halls and platforms. The problem is greatest for isolated ticket clerks late at night. See Poyner and Warne (1988) and Webb and Laycock (1992a).

Even the economy of crime was affected [by the eighteenth century boom in turnpikes]. The cash held by turnpike keepers proved a tempting target on the busy roads around London, and armed protection had to be provided on occasion.

P Langford *A Polite and Commercial People* (1989)

Where to find out more

Research sources

Poyner, B and Warne, C (1988)
Preventing Violence to Staff

Webb, B and Laycock, G (1992a)
Reducing Crime on the London Underground: an evaluation of three pilot projects

■

PUBLIC CAR PARKS

TYPICAL SECURITY ISSUES	
4	Cash robbery
11	Parked cars
	– theft of cars
	– theft of contents and components from cars
	– damage to cars
14	Fly-parking
19	Wilful damage
20	Graffiti

See the descriptions of each security issue in Part 2

Commentary

The *Secured Car Parks* guidelines published by the Association of Chief Police Officers (1992b) identify checklists of features which make car parks safer places. There are two checklists – for enclosed and outdoor car parks. However, the actual level of car crime depends on context and use as well as design features. Thus, a shopping mall car park which scored poorly on the checklist would probably have less crime than a railway station commuter car park with a better checklist score.

As well as car crime, the *Secured Car Parks* initiative is aimed at personal safety, an issue which is not covered in this guide. Although there is a risk of violent attacks including rape in car parks, the extent of this should not be over-emphasised. It may be that the main problem in car parks is the fear of attack in dark and deserted places.

CIRIA SPECIAL PUBLICATION 115 · DESIGN FOR INHERENT SECURITY · 1995

Contrary to the conventional view that women might suddenly be jumped upon or surprised by someone as they turn a corner, the more likely mode of attack would be by someone who had followed them into the car park rather than lying in wait for them. Such attacks would tend to be limited to multi-storey car parks with stair access. It is more difficult to follow someone unobserved in a lift and attacks are unlikely in a busy car park. Parking areas designed to reduce theft from cars through natural or formal surveillance are unlikely to be settings for violent attacks and rape.

Where to find out more

Research sources

See §11 'Parked cars' (page 77).

Good practice

Association of Chief Police Officers (1992b)
Secured Car Parks

■

PETROL AND SERVICE STATIONS

TYPICAL SECURITY ISSUES	
1	Common burglary
4	Cash robbery
5	Violence to staff
6	Shoplifting
10	Other theft from premises

See the descriptions of each security issue in Part 2

Where to find out more

Good practice

Lincolnshire, Nottinghamshire and Derbyshire Police (undated)
Garage Forecourt Security
Security for car sales lots.

Petrol Retailers Association (1993)
Petrol Filling Station Physical Security Measures

HOSPITALS

Commentary

The principal source of guidance on hospital security in Britain appears to be the *NHS Security Manual*, produced by the National Association of Health Authorities and Trusts (1992).

This guide is primarily concerned with the management of existing hospitals, not the design of new premises. It is wide-ranging, dealing with the security of equipment and the storage of medicines, linen, catering supplies, etc. It refers to the problems of cash and information security and also aggression and violence. Although it is not written for designers, they may find that this guide provides useful background information.

However, there is reason to doubt that conventional security practices in existing hospitals are satisfactory. A recent study showed that problems of loss prevention and personal security are a long way from being solved (Crime Prevention Consulting, 1993). The report also makes it clear that there are almost no reliable data about the nature and extent of crime and property loss within NHS hospitals. It must follow that without more information, it is impossible to judge the effectiveness of recommended security practices, and equally impossible to carry out reliable research studies on crime prevention and other security matters.

Research on risks

One study carried out in the mid-1980s did attempt to measure some of the crime problems in one hospital (Smith, 1987). It was a questionnaire survey of crime experienced by staff. What is interesting is that it focused on issues which are largely neglected in the *NHS Security Manual*. It showed that the highest risks for staff were loss or damage to their cars and bicycles, followed by the risk of theft of personal belongings. The threat of violence, which is dealt with to some extent in the *NHS Security Manual*, was also reported by Smith. However, there was some doubt about whether the problem was actually more serious for hospital staff than for the population as a whole.

The study by Crime Prevention Consulting confirmed the importance of the security issue 'Staff only access' (§16). It is probably more important in the hospital environment than anywhere else. The misuse of fire exits by staff wanting easy access is a common problem. Hospital consultants seemed to be the worst offenders. Examples are quoted where consultants to units like paediatric wards insisted on doors being left open until late at night to give them convenient access – thereby compromising the security of other staff, patients and property.

Design factors

On the question of the prevention of violence to staff in hospitals, there is a well developed view that the design and planning of accident and emergency departments can reduce the risk. A report by the Health and Safety Commission (1987) presented some design ideas, suggesting that the planning of reception and waiting areas, the provision of adequate personal space, the control of noise, and lighting and the use of colour, might contribute to the reduction of aggression. Such ideas will interest designers but it is unfortunate that little research has emerged to test their validity.

A recent research study identified two significant design factors for crime prevention in hospitals (NHS Estates, 1994). Firstly, internal security was improved in areas where there were *static* people (staff, patients waiting); the presence of *moving* people did not seem to improve security. Secondly, car crime outside hospital buildings was reduced in areas that were both overlooked by windows and close to a door; car crime did not seem to be reduced by overlooking unless there were also doors nearby.

Need for data and research

To summarise, there is a serious lack of good information about security issues within hospitals. It is clear that there are many uncertainties about the best approach to security in design. The problem of designing inherently secure hospitals is clearly a challenge to designers and their clients. It is also an important area for future research and development.

Where to find out more

Research sources

Crime Prevention Consulting (1993)
Preventing Crime in the NHS: the management challenge

Health and Safety Commission (1987)
Violence to Staff in the Health Services

Smith, L J F (1987)
Crime in Hospitals: diagnosis and prevention

Good practice

Burstein, H (1977)
Hospital Security Management

Colling, R L (1992)
Hospital Security
This American book is aimed mainly at security managers in existing hospitals.

National Association of Health Authorities and Trusts (1992)
NHS Security Manual

NHS Estates (1994)
Design Against Crime – a strategic approach to hospital planning

Nichols, J E (1983)
A Guide to Hospital Security

■

SCHOOLS

TYPICAL SECURITY ISSUES

1	Common burglary
5	Violence to staff
9	Personal belongings at work
10	Other theft from premises
11	Parked cars
12	Bicycle parking
16	Staff only access
19	Wilful damage
21	Arson

See the descriptions of each security issue in Part 2

Commentary

A recent survey indicated the relative frequency of different crimes in schools (Burrows et al, 1994). The survey collected data about crimes in 402 schools in the year 1991/92. The table (page 142) shows the percentages of the schools which suffered at least once in the year from various crimes (some schools suffered many times).

A research study carried out by Hope (1982) showed that general design characteristics of schools influence burglary risk. Using data from 59 schools in London he showed that older schools were much more secure than post-war schools. The average rate of burglary in schools which were large, modern and sprawling was over five times higher than in older schools which were compactly planned.

Issue	Crime	% of schools affected
1	Burglary	70%
	Attempted burglary	40%
5	Assault on staff	9%
8	Theft of personal belongings	71%
10	Theft of school property	41%
11	Theft of or from staff vehicles	22%
19/20	Deliberate damage	67%
21	Arson	19%

The design characteristics which correlated most with burglary were:

- the size of buildings and the site
  (larger schools and sites – more burglary)

- the number of separate buildings
  (more buildings – more burglary)

- the lack of compactness in the design
  (sprawling layout – more burglary)

- the use of landscaping – trees, shrubs, etc
  (more landscaping – more burglary).

All of these characteristics effect the extent to which surveillance of the schools was possible. All but two of the schools surveyed had resident caretakers and it was noted that most of the break-ins occurred at points out of sight from the caretakers' accommodation.

Another characteristic which contributed to the security of the older schools was that they were surrounded by high perimeter walls or railings. Post-war schools had more open boundaries, often with low walls at the front.

Compact schools in housing areas usually have lower levels of crime than more isolated schools on larger sites.

Security lighting

Among current proposals to improve the security of schools is the idea that lighting will deter intruders who may be intent on burglary, vandalism or arson (Department for Education, 1993). No evaluation study of school lighting is known, even though it would be straightforward to assess.

Logically, lighting could only be of value as an aid to formal or natural surveillance from caretakers or surrounding streets, residents or other activities. It seems unlikely that lighting could enhance security in a school which was not inherently secure.

See also the discussion in 'Lighting and security' in Part 4 (page 158).

Where to find out more

The common practice of hanging up caps, cloaks and bonnets along the walls of the school-room would appear to be the result of cheeseparing economy... On the other hand experience in some of the denser and rougher parts of London shows that when an outer corridor or entrance is used as a cloak-room many of the articles are stolen.

E R Robson *School Architecture* (1874)

Research sources

Atkins, S, Husain, S and Storey, A (1991)
The Influence of Street Lighting on Crime and Fear of Crime

Burrows, J, Shapland, J, Wiles, P and Leitner, M (1994)
Arson in Schools

Hope, T (1982)
Burglary in Schools: the prospect for prevention

Ramsey, M (1991)
The Effect of Better Street Lighting on Crime and Fear of Crime: a review

Good practice

Cheetham, D W (1994)
Dealing with Vandalism: a guide to the control of vandalism

Department of Education and Science (1987)
Crime Prevention in Schools: practical guidance

Department of Education and Science (1989a)
Graffiti Removal and Control

Department of Education and Science (1989b)
Crime Prevention in Schools: specification, installation and maintenance of intruder alarm systems

Department for Education (1993)
Crime Prevention in Schools: security lighting

Williams, D (1989)
'Lessons in security' and 'Post-vandalism'
Descriptions of recent schools designed to combat crime.

Zeisel, J (1976)
*Stopping School Property Damage: design and administrative
guidelines to reduce school vandalism*

■

CIRIA SPECIAL PUBLICATION 115 · DESIGN FOR INHERENT SECURITY · 1995

OTHER PUBLIC BUILDINGS

TYPICAL SECURITY ISSUES	
1	Common burglary
8	Customer belongings
10	Other theft from premises
11	Parked cars
15	Customer only access
16	Staff only access
19	Wilful damage
20	Graffiti

See the descriptions of each security issue in Part 2

Where to find out more

Good practice

East Anglian Church Security Group (1992)
Churches: A Crime Prevention Strategy
A collaboration between Norfolk, Suffolk, Essex and Cambridgeshire police crime prevention officers.

Museums and Galleries Commission
The Commission issues guidance notes on security.

Sussex Police (undated)
Security of Police Stations

■

PART 4

NOTES ON SECURITY MEASURES

This guide is primarily concerned with design for inherent security, not the specification of security measures. Security measures are well covered in British Standards and other guidance literature and the following notes are not intended to replace or improve on this technical data. Here the purpose is to summarise and review some key topics from the existing literature, particularly where security measures have a bearing on inherent security.

There are five sections in Part 4. The first two discuss windows and doors – very much a focus of attention in the existing literature. The third section looks at 'natural ladders' – a term used to describe architectural or landscape elements that give unintended assistance to criminals attempting to gain illegal access.

The last two sections are somewhat different. The section on lighting and security reviews the research evidence for lighting as a means of crime prevention. The last section discusses the feasibility of using security categories to classify risks and security measures.

WINDOWS AND BURGLARY

Security design has traditionally devoted a great deal of attention to windows. They are a weak point in the building enclosure and can provide a vulnerable route for break-ins. Design decisions about windows are important for inherent security.

Windows and security

Good inherent security reduces the vulnerability of buildings to burglary, but it is still necessary to provide adequate protection against windows being used as a route for break-ins. The principles of inherent security suggest three factors that are relevant to the specification of security measures for windows:

- not all windows are potential routes for break-ins (for example, high level windows are less vulnerable)

- site layout and building design can reduce the number of windows that are potential routes for break-ins

- windows in the right place can contribute to inherent security by allowing surveillance

Where window strengthening is required, existing security design guides and manufacturers' datasheets provide copious information.

Commentary on existing guidance

The following sections summarise selected items of advice about windows from British Standards and other existing design guidance. Many design guidance publications go into much greater depth and should be referred to for further detail.

Window location and vulnerability

Windows are only a security risk if they offer a credible entry route, so inaccessible high-level windows are not of concern for security. Where possible, it is desirable to avoid windows in vulnerable locations.

➤ Ground level windows and windows accessible from 'natural ladders' like roofs, pipes, trees and so on, are potentially at risk.

➤ Windows in locations which lack surveillance are more vulnerable.

➤ Roof glazing or roof windows are just as vulnerable as conventional windows, if they are accessible.

Window opening dimensions

The aperture was so small, that the inmates had probably not thought it worth while to defend it more securely; but it was large enough to admit a boy of Oliver's size.

Dickens *Oliver Twist* (1839) – from Chapter 22, 'The Burglary'

➤ To prevent people getting though a window opening, the largest dimension should not exceed 125mm. A maximum area of 0.05m^2 is also recommended and it is estimated that an opening of 200×280mm (0.056m^2) is large enough to provide access.

➤ A much smaller opening is sufficient for someone to reach through and release an opening catch.

➤ Security is compromised when windows are left open, but this may be necessary for ventilation. Window restrictors are sometimes proposed, to give a maximum opening of 100 or 125mm. A restrictor is useless if it can be forced or released easily.

Window construction

➤ Ensure that windows are fixed securely in the wall, otherwise a whole window could be removed.

➤ If window fixings or even hinges are accessible from outside they may be tampered with. For example, the pins in projecting hinges can be drilled out with battery-operated hand tools.

➤ If there are external fixings, use ones which are tamper-proof.

Glazing

Glass is often broken to gain access to buildings. However, guidance normally assumes that thieves do not like entering through broken glazing because of noise and the risk of injury. More often they reach through the broken pane to release an opening light. This implies that glazing is less vulnerable in fixed lights or in opening lights which cannot be released after breaking the glass.

➤ In vulnerable situations use laminated glass; even double-glazing or double-glazed units may deter some attackers. If security glazing is to act as a deterrent it must be obvious that it is not ordinary glass.

➤ Leaded lights are particularly weak and can often be penetrated without even smashing the glass.

➤ Plastic glazing materials can be flexible and may be sprung out of the frame.

➤ Where windows are divided into small panes it is easier to smash one pane to release an internal catch.

➤ If glass is not strongly fixed an entire pane may be removed.

➤ If ventilation is provided mechanically or by means of grilles, windows can be kept closed or fixed glazing can be used.

➤ Low-level glazing in ground level windows is sometimes reckoned to be at risk from being kicked in.

Shutters and bars

Bars are the traditional way of preventing people getting through windows or openings. They can be decorative but are more often ugly and intrusive, particularly if fitted as an afterthought. If bars are really necessary it is desirable that they are treated as an integral part of a design.

➤ Bars or grilles should be fitted inside the window. External bars and grilles are more vulnerable and need particularly secure fixings; however, they can be more decorative.

➤ External bars require inward-opening windows.

➤ Opening grilles, or bars with opening panels, are more vulnerable than permanently fixed ones.

➤ The strength of the bars and fixings should be adequate: there is no point installing flimsy bars or grilles.

➤ Solid shutters can hide valuable goods from the eyes of potential thieves. Screening curtains, blinds or even

Bars are familar in traditional architecture, often well designed and avoiding the negative image of some present-day security devices.

obscured glazing can achieve the same effect, which only applies in locations which offer a view to potential thieves.

Window ironmongery and locks

➤ Strong window catches reduce the risk of windows being forced open. Multi-point or espagnolette-type ironmongery also makes forcing more difficult.

➤ Locking catches are often recommended. The fact that a window catch is lockable does not help it to resist forcing – that depends on the strength of the catch. The purpose of window locks is to prevent a window being opened by someone who has broken the glass or forced a small light. This requires key-operated locks, not catches with snibs or turnbuttons.

➤ To be effective as a deterrent, robust ironmongery or lockable catches must be visible to potential attackers.

Where to find out more

Good practice

British Standard 8220: Part 2: 1987
Security of Buildings against Crime: Offices and Shops

British Standard 8220: Part 3: 1990
Security of Buildings against Crime: Warehouses and Distribution Units

Cumming, N (1992)
Security: a guide to security system design and equipment selection and installation

Fire Protection Association (1985)
Security Precautions: windows and rooflights

Underwood, G (1984)
The Security of Buildings

SECURITY OF DOORS

The conflict between security and other design requirements is most clearly evident for doors. How do you allow some people to get in and out of a building, but not everybody?

Doors and security

A door at the back of an electrical superstore is made of steel, has no external ironmongery, and is protected by anti-ramraid bollards. Could a door in this location have been avoided altogether?

Like windows (page 149), doors are a weak point in the building enclosure and are generally regarded as being vulnerable to break-ins.

The principles of inherent security indicate that the risk of break-ins is minimised if doors are located only where they are subject to formal or informal surveillance. Without surveillance, almost any door can be broken down by determined criminals.

Break-ins are not the only problem. Intruders may gain entry surreptitiously through doors which are open for the use of authorised personnel or legitimate visitors. Effective control or supervision is required to prevent this (see §15 'Customer only access' and §16 'Staff only access').

A conflict between fire escape and security is often put forward – this topic is discussed in §1 'Common burglary' (see page 30). Fire doors must be openable from inside while the accommodation they serve is occupied, but external ironmongery can be omitted to prevent casual entry. As well as offering a potential access route, fire doors may be used as an escape route after theft or some other crime: if they are not supervised it is possible to fit alarms that are activated when the door is opened, but there must be a rapid response to the alarm signals.

Commentary on existing design guidance

A door's ability to resist break-ins depends on its physical strength, the way it opens, and the ironmongery.

Door construction

➤ Panels in panelled doors are often weak, as are areas of glazing, particularly at low level. Glazing in doors is normally in safety glass to prevent injury from breakages, but additional strength may be specified for security.

➤ Hollow-core doors are very weak indeed and should never be used if security is a concern.

➤ Ensure that door frames are as strong as the locks and keeps fixed to them, and ensure that the method of fixing into the wall is adequate. Note that excessive morticing of doors and frame for locks and other ironmongery can weaken them – sections should be sufficient for the intended ironmongery.

➤ Steel doors or timber doors faced with steel sheet can be extremely robust. If the steel facing masks the frame openings the door is more difficult to force. Steel shields can be used locally to protect hinges and locking points.

➤ Small glazed panels or peepholes can contribute to access control in a building where doors are usually locked – a maximum width of 60mm is recommended.

Door configuration

➤ Double doors are usually weaker than single doors.

➤ Inward-opening doors are usually weaker than outward-opening doors, as they rely solely on the lock and/or bolts to combat a 'battering ram' attack. Outward opening doors are vulnerable to being pulled open – some design guides suggest avoiding anything that could act as an anchor for a tow-rope.

➤ Sliding doors and pivot doors are harder to secure against break-ins than hinged doors.

➤ Revolving doors can be a deterrent – it is more difficult for a criminal to make a rapid escape if challenged.

➤ Where outward-opening doors discharge into public areas, it is better to install safety rails than to recess the doors – recesses create hiding places and obstruct surveillance of the door.

Recessed doors can invite undesirable activities.

Ironmongery and locks

➤ Locks with sophisticated keying and multiple key differs resist unauthorised release. Excellent products are readily available but designers and specifiers should not overlook other aspects of vulnerability.

➤ Thieves normally prefer to open a door rather than smash it. Locks which allow the door to be freely opened from the inside make this easier, but deadlocks can conflict with fire escape requirements.

➤ Lever handles may be easier to attack than knobs.

➤ External strap hinges are particularly easy to attack, and any projecting hinges are vulnerable. Plastic and aluminium hinges are weaker than equivalents in steel or brass. Hinge bolts can counteract these weaknesses, and stengthen a door generally.

➤ Surface-fixed locks are usually weaker than locks that are morticed into a door.

The guard was inspected ... and His Excellency then came up the stairway to the grand portico... At a given signal the doors were opened – no key as there was no lock – and Lady Irwin and His Excellency went in.

Sir Edwin Lutyens on the opening of the Viceroy's House [ie. Palace] in New Delhi (1929)

Where to find out more

Good practice

British Standard 8220: Part 2: 1987
Security of Buildings against Crime: Offices and Shops

British Standard 8220: Part 3: 1990
Security of Buildings against Crime: Warehouses and Distribution Units

Cumming, N (1992)
Security: a guide to security system design and equipment selection and installation

Fire Protection Association (1985)
Security Precautions: Doors

Underwood, G (1984)
The Security of Buildings

■

NATURAL LADDERS

The preoccupation with the prevention of burglary in existing design guidance has led to a great deal of advice about how to stop people scaling barriers or climbing where they are not supposed to. Features that help climbing are often called 'natural ladders'.

The threat presented by 'natural ladders'

A 'natural ladder' is a feature on a building or in the environment that provides the opportunity for climbing, even though it was not designed for that purpose. Natural ladders can be considered part of the inherent security (or insecurity) of a design – they are not part of a purpose-designed security system.

Natural ladders are a security risk if they could help offenders gain illegal entry to buildings or damage them. This could happen in two ways: firstly, by allowing offenders to scale boundary fences or walls; and secondly, by helping them reach vulnerable parts of the building enclosure. The preventive strategy is self-evident: avoid natural ladders which could help offenders in these ways.

Natural ladders are primarily relevant to burglary (§1) and damage to buildings (§19, 20, 21); they have less importance for most of the other security issues in Part 2.

Various examples of natural ladders are identified in the existing guidance literature, but every designer will be able to imagine an almost infinite number of possible natural ladders. Some design guidance calls for the elimination of *all* design features which might provide natural ladders. However, the real test is whether a design feature which could be climbed gives access to a weak point. For example, an external rainwater pipe on a blank facade may

Both the adjacent footbridge and the stacked materials could act as natural ladders to assist scaling of this perimeter fence.

be a minimal security risk, but if it passed close to a window with a broad sill it could be a severe security risk.

Commentary on existing design guidance

Perimeter fences

➤ Do not plant trees in positions where, when they are mature, branches will spread over a perimeter fence.

➤ In external storage areas do not stack goods close to a perimeter fence.

➤ Place lamp-posts inside fences and use long outreach arms, rather placing them outside a fence but close to it.

➤ Corners in perimeter fences should never be less than 135° to prevent bridging across the corner.

➤ Use barriers to keep vehicles at least 2m away from a perimeter fence, in case the vehicle is used to scale the fence.

Building envelope

➤ Keep buildings at least 3m from perimeter fences to prevent the gap being bridged.

➤ Avoid external rainwater pipes or drainage pipes whenever possible. If they are unavoidable, use square pipes fixed tight to the wall surface.

➤ Protect roof windows as securely as ground floor windows.

➤ Do not plant trees in positions where, when they are mature, branches will spread over roofs or come close to windows.

➤ Use overhanging eaves and eliminate any features which would help climbing onto the roof.

➤ Avoid trelliswork on a building facade, especially if it is strongly constructed.

➤ Low pitched roofs, especially flat roofs, can often give access to upper floor windows or to higher roofs.

➤ External landscape features, like mounds or planting boxes, can reduce the effective height of barriers or roofs.

➤ Avoid projecting sills, stringcourses or other features on the building facade.

Where to find out more

Practically all security design publications draw attention to various examples of natural ladders, but they are not treated systematically as a distinct topic.

■

LIGHTING AND SECURITY

The use of lighting as a means of providing security is well established in much crime prevention literature and in the lighting industry's own promotional literature (Electricity Council, 1988). The purpose of this note is to summarise what research has to tell us about lighting and crime and to assess the relevance of lighting to the security issues covered in this guide.

Research on lighting and crime

Evaluation research on the effect of lighting on crime has focused on street lighting. Much effort has been expended by Home Office funded researchers to assess the impact of improving street lighting (Ramsey, 1991; Atkins et al, 1991). The general conclusion is that increasing the levels of illumination in city streets has no clearly measurable effect on crime, but it can reduce anxiety or fear of crime, particularly for women. This conclusion is in line with earlier evaluation work in the United States (Tien et al, 1979).

The findings from several studies by Kate Painter are also worth noting (Painter, 1988, 1989, 1991 and 1994). Evaluations were made of the effect of significantly upgrading the lighting of poorly lit side streets in three locations in the inner London suburbs. The streets chosen for the experiments were all useful pedestrian routes which were particularly dark at night (and which had little vehicular traffic). Each route passed through some narrower restriction – railway bridges in two cases.

Painter's research was based on interviews with people using these streets, both before and six weeks after new lighting was installed. Although the findings were positive, the very small numbers of crimes reported and the shortness of the time period make it difficult to be certain that there was a true reduction in crime (see also comments in Ramsay, 1991). However, in the third study there is clear evidence that the introduction of lighting in what was a virtually unlit side street dramatically reduced 'incivilities' (Painter, 1991). In this case there had been a problem of men using the street at night as a place to urinate – an incivility which upset elderly female residents. The new lighting appears to have eliminated this problem. Before and after maps of incident locations show very clear differences (Painter, 1991, pp 82-84).

These studies also showed that pedestrian usage in the streets increased following the lighting improvement, by between 34% and 101% (Painter, 1994). The increases applied to both men and women. Although most interview surveys suggest that women are much more concerned than men about being attacked in badly lit areas, this objectively counted data reveals that improved lighting encourages use by men as well as women.

For the designer and building owner there seem to be important implications from Painter's research. Good illumination of an area discourages some kinds of nuisance behaviour and may increase usage at night. Both of these effects will generally be considered beneficial for business by building owners or tenants. Furthermore, the presence of more people may provide better natural surveillance and thereby reduce crime in general.

There is little evaluation literature on lighting and security in commercial settings, but two American sources provide some evidence that good lighting may aid surveillance at night. The first is an evaluation of a crime prevention project in Portland, Oregon (Griswold, 1992). The combination of a police campaign to increase the physical security of shop premises together with the upgrading of street lighting along a major commercial corridor, was found to reduce burglary. The second source reviews a number of studies of robbery prevention measures for convenience stores (Hunter and Jeffrey, 1992). The lighting of entrances, and possibly parking lots, was considered to be a factor in deterring attacks on stores open late at night.

Role of lighting in this guide

It is clear from the research on street lighting that it would be wrong to think that by simply increasing the level of illumination crime will decrease in a dramatic way. Nevertheless, there is evidence which suggests that lighting can have an effect on behaviour, particularly where lighting is introduced in poorly lit places.

The functions of lighting in relation to the security of non-residential buildings which arise from research might be summarised as follows.

- *Facilitates informal surveillance by the public*
  Lighting an area at night enables people who might report or even intervene in criminal activity to see suspicious behaviour and take action. A potential criminal might be deterred if he considers that lighting increases the risk of being caught.

- *Increases use by the public, which further increases surveillance*
  This might apply to a route through or alongside a building complex.

- *Facilitates surveillance by employees*
  Where staff are employed in the hours of darkness, lighting may help them to keep watch, identify potential problems and take preventive action. The risk of being seen by staff may deter potential offenders.

All three of these security functions of lighting can be exploited by designers to reduce the risks of crime. However, if we examine the list of 21 security issues in Part 2 of this guide (see pages 22-23), not all will benefit from the introduction of increased illumination or special security lighting.

Most of the security issues only apply in situations which are reasonably well lit. For example, the issues relating to robbery (§4), violence to staff (§5), shoplifting (§6), customer only access (§15) and loitering youths (§18) all take place in environments which will already be well lit for the operation the businesses concerned.

A similar argument applies to staff pilferage (§7), safety of customers belongings (§8), security for personal belongings (§9) and other thefts from premises (§10). Areas around loading bays (§13) would normally be lit if they are used after dark. Providing additional lighting in situations which are already well lit seems unlikely to enhance the security aspects of the design.

There are some other issues where it is unclear whether lighting would have any benefit. For example, smash and grab raids (§2), ramraiding (§3) and fly-parking (§14) are just as likely to occur in well-lit and poorly-lit areas.

It is unclear how lighting affects property damage (§19), graffiti (§20) and fly-tipping (§14). Damage and graffiti may even be promoted by lighting if the aim of the vandals is to draw attention to their activities. Lighting may have little effect on fly-tipping as it only occurs in unsupervised places, and might just as well be done in daylight.

This analysis leaves us with only five security issues in which lighting could play a significant part in crime prevention. The first is common burglary (§1). The section on this issue draws attention to the value of surveillance in controlling common burglary and suggests lighting the facades of buildings which can benefit from surveillance, in particular facades that face well-used streets. There does not seem to be a need for very strong levels of illumination and it is probable that most current street lighting is sufficient. There seems little point in providing lighting in areas where no surveillance is possible, but judgement has to be left to the designer.

A particular building type where lighting might be used to reduce burglary risk is schools. Schools often have large perimeters which offer many opportunities for surveillance from neighbouring houses as well as resident caretakers. Under these circumstances the expense of security lighting may be considered acceptable, although no evaluation of the effect on burglary has been made. The matter is discussed in more detail in a Department for Education Building Bulletin (Department for Education, 1993).

In relation to car parking (§11), areas which are in use after dark will benefit from lighting, but its value must be primarily for the safety and convenience of users rather than for crime prevention or fear reduction. Also, lighting by itself will not prevent crime; there must be a capability for surveillance. For example, there is no reason to expect that good lighting would make much difference in a little used, isolated car park at night.

There has been no specific research into various levels of illumination in car parks and the effects on fear and crime, but it is likely that good distribution of light is more important than high intensity. It is common experience that the general level of illumination in car parks in many towns and cities is poor. The suggestion in the police *Secured Car Parks* initiative that walls and ceilings should be in a light colour seems to make sense and contrasts strongly with the widespread tendency to leave exposed concrete structures to get progressively dirtier and darker (Association of Chief Police Officers, 1992b).

The section on safe bicycle parking (§12) makes no specific recommendation for the lighting of cycle stands or cycle parks, although this would certainly assist surveillance. Much depends on the situation the designer is considering. It is assumed that if cycle stands are in well-used public areas, as recommended, they will be adequately lit in the evening.

Where separate entrances are provided for staff only access (§16) there will normally be lighting for the convenience and safety of users. However, where there is likely to be a more serious security problem, such as a risk that staff may be attacked on entry or when leaving, more care will be needed to light the area to facilitate surveillance from inside the building and by staff approaching the entrance.

The remaining security issue is arson (§21). Here the reason for lighting would be to facilitate surveillance, in order to discourage access by children and youths. The situation is, perhaps, most relevant to schools, for which a guide on security lighting has been published, as mentioned above (Department for Education, 1993). Lighting which reduces the risk of arson would also reduce the risk of burglary (§1) – and vice versa.

Design problems with security lighting

As a footnote, it might be said that many current installations of security lighting seem to emphasise intensity of illumination over quality. This may be due to over enthusiasm for security lighting, or attempts to compensate for poorly chosen or located luminaires.

The design of security lighting must give careful consideration to its lighting function. For example, wall mounted luminaires are commonly fixed on the external face of a building. This works well if surveillance is from the inside looking out, but when viewed from outside the result may be disabling glare, which makes surveillance impossible.

Another common problem is that security lighting can give rise to lighting pollution (Heywood, 1994). The use of high level and high intensity lighting in large industrial yards or car parks next to housing or other areas with low illumination can create annoyance and disamenity. High levels of illumination in one place may also increase the vulnerability of neighbouring properties by making them appear particularly dark and less likely to benefit from surveillance by neighbours, passing pedestrians and traffic.

Clearly, the design of lighting for security has to keep all these problems in mind.

Where to find out more

Research sources

Atkins, S, Husain, S and Storey, A (1991)
The Influence of Street Lighting on Crime and Fear of Crime

Griswold, D B (1992)
'Crime prevention and commercial burglary'

Heywood, K (1994)
Light Pollution

Hunter, R D and Jeffrey, C R (1992)
'Preventing convenience store robbery through environmental design'

Painter, K (1988)
Lighting and Crime Prevention: the Edmonton Project

Painter, K (1989)
Lighting and Crime Prevention for Community Safety: the Tower Hamlets Study

Painter, K (1991)
An Evaluation of Public Lighting as a Crime Prevention Strategy with Special Focus on Women and Elderly People

Painter, K (1994)
'The impact of street lighting on crime, fear and pedestrian street use'

Ramsey, M (1991)
The Effect of Better Street Lighting on Crime and Fear: a review

Tien, J, O'Donnell, V F, Barnett, A and Mirchandani, P B (1979)
Street Lighting Projects: National Evaluation Program

Good practice

Association of Chief Police Officers (1992b)
Secured Car Parks

Chartered Institute of Building Services Engineers (1991)
Security Engineering

Department for Education (1993)
Crime Prevention in Schools: security lighting

Electricity Council (1988)
Essentials of Security Lighting: commercial and industrial premises

Fire Protection Association (1985)
Security Precautions: external security lighting

Lyons, S (1992)
Lighting for Industry and Security
Chapter 12 deals with 'Interior security lighting' and chapter 17 with 'Exterior security lighting'.

■

THE PROBLEM OF SECURITY CLASSIFICATION

Building owners and designers have to decide what level of security protection they need, and how to provide it. Rather than treating every building as a unique problem, it seems convenient to establish security categories. How are security categories defined and how are they used?

The designer's dilemma

A security design guide for architects states that, 'a decision on the degree of protection required is an essential starting point' (England, 1992). Unfortunately this proposition is not followed up with any data about the range of possible 'degrees of protection', or how to identify the right one for a given project. So the designer is left with hardly any guidance on what he has been told is an 'essential starting point'.

Similarly, manufacturers' datasheets describe their products in terms which are left undefined; for example, a typical manufacturer offers door locks of three standards, to 'cover medium security, high security and maximum security requirements'. This is imprecise and does little to help the designer specify the right equipment for a particular project.

What is the 'level of security'?

Many problems arise because 'level of security' is used to mean different things at different times. The term describes at least three distinct concepts:

1   **threat** – a measure of the crime threat to a building resulting from its location and use

2   **protection** – a description of the preventive strategies and measures that are adopted

3   **exposure** – the overall security risk that a building faces, taking account of both threats and protection.

Thus, the designer of an electrical superstore might say that the project presents a severe security problem (high level of threat); the building specification might call for high security shutters to the display windows (high level of protection); and if, through good design and management, the building and its contents do not suffer from crime, a high level of security has been achieved (*low* level of exposure).

In many situations high levels of security (that is, low exposure to crime) can be achieved without elaborate, high security protection – and where possible, this must be the most desirable outcome.

However, security design manuals and manufacturers' datasheets concentrate almost exclusively on how to increase protection. Their guidance is dominated by measures to combat one particular threat – illegal entry. But there are many other security issues, as discussed in Part 2.

Classification schemes

When a precise and quantified analysis of security is attempted, things get difficult. Many proposed classification schemes have been put forward, but it seems fair to say that there is no widespread agreement amongst specialists.

Most proposals concentrate on classifying the level of protection offered by building components – doors, windows, locks and so on – by defining a series of tests that the components must be subjected to: their security classification depends on performance in these tests. The tests are designed to simulate the kinds of attack that would be made by would-be intruders.

A European standards committee, for example, has developed a proposal for six 'security rating classes' (Loss Prevention Council, 1993). Tests of three types are defined (in detail): soft body impacts, static loads, and manual tests with tools (a different toolkit and period of resistance is specified for each rating class).

The proposal does go on to suggest what security class is appropriate for different building types in the UK. But whereas the tests are set out in detail, the list of building types is extremely sketchy.

This one-sided preoccupation with classifying the protection of components is of limited value to the architect and building owner, whose concern is with overall exposure to crime. They want a method which matches the level of protection to the level of threat.

Points systems

An alternative to a set of security categories is a points system. One was developed by the Chartered Institute of Building Services Engineers (1991).

There are many factors contributing to a building's overall level of exposure to crime, and a points system aims to assess the factors on a building-by-building basis. The CIBSE system describes buildings with respect to 21 attributes – 13 describe the level of threat and eight the level of protection. The total score for a building indicates the level of exposure to crime.

A difficulty with any points system is that it may be possible to achieve a good overall score despite some very weak features; the system has to be used intelligently to produce worthwhile results.

The CIBSE points system, like other classification systems, concentrates on burglary and has less relevance to other security issues.

Summary

Although it is an attractive idea to develop security classification systems, the task turns out to be extremely difficult, and it has not yet been successfully achieved. Perhaps with further research and development it will be. Meanwhile designers and building owners must always check proposed security measures for their relevance and usefulness in each particular project.

Where to find out more

Research sources

Loss Prevention Council (1993)
Specification for Testing and Classifying the Burglary Resistance of Building Components

Good practice

Chartered Institute of Building Services Engineers (1991)
Security Engineering

England, N (1992)
'Security equipment'

■

PART 5   **FURTHER INFORMATION**

ORGANISATIONS TO CONTACT

The following list includes national bodies who may be helpful to designers and building owners or users. Organisations whose role is specific to particular security issues or building types are listed in the relevant sections earlier in the guide.

Arson Prevention Bureau
140 Aldersgate Street, London EC1A 4DD
tel 0171-600 1695 fax 0171-600 1487

Established by the insurance industry and the government to coordinate a national programme for action against arson. Publishes review papers and datasheets.

British Hardware and Housewares Manufacturers' Association
Brooke House, 4 The Lakes, Bedford Road, Northampton NN4 0YD
tel 01604 22023 fax 01604 31252

Represents manufacturers whose product ranges include security devices.

British Security Industry Association
Security House, Barbourne Road, Worcester WR1 1RS
tel 01905 21464 fax 01905 613625

Publishes a members' list of firms involved in the security industry.

Building Research Establishment
Garston, Watford, Herts WD2 7JR
tel 01923 894040 fax 01923 664010

A BRE research team is active in security research and testing.

Chartered Institution of Building Services Engineers
Delta House, 222 Balham High Street, London SW12 9BS
tel 0181-675 5211 fax 0181-675 5449

Building services engineers are responsible for the specification of electronic security equipment. Publishes a manual on security.

Crime Concern
Signal Point, Station Road, Swindon, Wilts SN1 1FE
tel 01793 514596 fax 01793 514654
Crime Concern Scotland
29 Frederick Street, Edinburgh EH2 2ND
tel 0131-226 5661 fax 0131-226 5662

National crime prevention organisation. Provides advice and
consultancy services to local authorities and public sector
organisations, as well as issuing guidance publications. Main
emphasis is on social prevention.

Glass and Glazing Federation
44-48 Borough High Street, London SE1 1XB
tel 0171-403 7177 fax 0171-357 7458

An industry association concerned with all aspects of glass and glazing,
including security. Publishes advisory datasheets.

Home Office Crime Prevention Unit
Home Office, 50 Queen Anne's Gate, London SW1H 9AT
tel 0171-273 2764

Initiates and publishes Home Office research on crime and crime
prevention.

Home Office Crime Prevention Centre
Cannock Road, Stafford ST17 0QG
tel 01785 58217 fax 01785 232524

Police organisation that coordinates policy on crime prevention
through architectural and environmental design, and trains police
architectural liaison officers.

Laminated Glass Information Centre
299 Oxford Street, London W1R 1LA
tel 0171-499 1720 0171-495 1106

Trade organisation promoting and providing information about
laminated glass. Sponsored by manufacturer of laminated glass
interlayer material.

Loss Prevention Council
140 Aldersgate Street, London EC1A 4HX
tel 0171-606 1050 fax 0171-600 1487

The insurance industry's technical organisation. Publishes a series of
security precautions datasheets. UK representative on European
security standards committees.

■

REFERENCES AND BIBLIOGRAPHY

The following list includes all the publications referenced earlier in the guide, plus some general references that are relevant to design and security. Those references which are recommended for further reading or as additions to a reference library are highlighted in boxes, and some have brief annotations.

Arson Prevention Bureau (1992)
Prevention and Control of Arson in Industrial and Commercial Premises
APB (London): Arson Update 1/92

Arson Prevention Bureau (1993)
How to Combat Arson in Schools
APB (London)

Association of British Insurers (1991a)
Guidelines on Shop Front Protection
ABI (London): Crime Prevention Technical Unit Information Bulletin G/220/710

Association of British Insurers (1991b)
Impact Attacks on Commercial and Industrial Premises using Vehicles
ABI (London): Crime Prevention Technical Unit Information Bulletin P/431/710

Association of Chief Police Officers (1992a)
Secured by Design: Commercial
The basis of the advice given by police for commercial developments. The focus is on design aspects rather than hardware and security systems. Available from local police forces (ask for the Architectural Liaison Officer).

Association of Chief Police Officers (1992b)
Secured Car Parks
The police criteria for assessing car parks. Successful designs are
allocated silver or gold awards. There are two checklists of criteria, for
open and enclosed car parks. Available from local police forces (ask for
the Architectural Liaison Officer).

Atkins, S, Husain, S and Storey, A (1991)
The Influence of Street Lighting on Crime and Fear of Crime
Home Office (London): Crime Prevention Unit Paper 28

Austin, C (1988)
The Prevention of Robbery at Building Society Branches
Home Office (London): Crime Prevention Unit Paper 14

Barefoot, J K (1990)
Employee Theft Protection
Butterworths (Boston, Mass)

Bennett, T and Wright, R (1984)
Burglars on Burglary: prevention and the offender
Gower (Aldershot, Hants)
A book by criminologists, not designers.

Brantingham, P L, Brantingham, P J and Wong, P (1990)
'Malls and crime: a first look'
Security Journal vol 1, pp 175-181

British Council for Offices
The BCO publishes six datasheets on: office crime, bombs and
terrorism, office car parking, CCTV, disaster recovery, business parks.
Contact data: British Council for Offices, College of Estate
Management, Whiteknights, Reading RG6 2AW tel 01734 885505.

British Standard 5051: Part 1: 1973
Bullet-resistant Glazing
BSI (Milton Keynes)

British Standard 5544: 1978
Anti-Bandit Glazing (glazing resistant to manual attacks)
BSI (Milton Keynes)

British Standard 8220: Part 2: 1987
Security of Buildings against Crime: Offices and Shops
BSI (Milton Keynes)
The recommendations concentrate on the specification of doors,
windows and rooflights.

British Standard 8220: Part 3: 1990
Security of Buildings against Crime: Warehouses and Distribution Units
BSI (Milton Keynes)
This part of the standard gives recommendations for the specification of doors, windows and rooflights and discusses security management and alarm systems. Particular attention is given to shutter and door types for use in warehouses. A revised edition is forthcoming.

Buckle, A and Farringdon, D P (1984)
'An observational study of shoplifting'
British Journal of Criminology vol 24, pp 63-73

Buckle, A, Farringdon, D P, Burrows, J, Speed, M and Burns-Howell, A (1992)
'Measuring shoplifting by repeated systematic counting'
Security Journal vol 3, pp 137-146

Burrows, J, Shapland, J, Wiles, P and Leitner, M (1993)
Arson in Schools
Arson Prevention Bureau (London)

Burrows, J and Speed, M (1994)
Retail Crime Costs 1992/93
British Retail Consortium (London)

Burstein, H (1977)
Hospital Security Management
Praeger (New York)

CFPA Europe (Conference of Fire Protection Associations) (1989)
Arson Dossier (UK edition)
Fire Protection Association (London)

Chartered Institute of Building Services Engineers (1991)
Security Engineering
CIBSE (London): Applications Manual 4
This guide is written for services engineers but many sections are accessible to a wider audience. It begins with a points system for assessing security in particular projects, and then discusses most types of electronic security system. There are notes on security in various building types.

Cheetham, D W (1994)
Dealing with Vandalism: a guide to the control of vandalism
CIRIA (London): Special Publication 91

Clarke, R V (ed) (1978)
Tackling Vandalism
Home Office (London): Home Office Research Study 47

Clarke, R V (ed) (1992)
Situational Crime Prevention: successful case studies
Harrow & Heston (Albany, NY)

Clarke, R V and Harris, P M (1992)
'A rational choice perspective on the targets of automobile theft'
Criminal Behaviour and Mental Health vol 2, pp 25-42

Clarke, R V and Mayhew, P (1980)
Designing out Crime
HMSO (London)

Colling, R L (1992)
Hospital Security (3rd edn)
Butterworth-Heinemann (Boston, Mass)

Crime Prevention Consulting (1993)
Preventing Crime in the NHS: the management challenge
CPC (London)
Contact data: Crime Prevention Consulting, 47c Marloes Road,
London W8 6LA.

Crowe, T D (1991)
Crime Prevention through Environmental Design
Butterworth-Heinemann (Boston, Mass)
Although its research credentials are uncertain, this book represents an
important statement of the CPTED approach to security design which
grew out of US research and theorising. Lots of case studies and ideas
about planning for security.

Cumming, N (1992)
*Security: a guide to security system design and equipment selection and
installation* (2nd edn)
Butterworth-Heinemann (London)
A comprehensive guide to the design and specification of security
equipment for dealing with intruders.

Department of Education and Science (1987)
Crime Prevention in Schools: practical guidance
HMSO (London): DES Building Bulletin 67

Department of Education and Science (1989a)
Graffiti Removal and Control
HMSO (London): DES Design Note 48

Department of Education and Science (1989b)
*Crime Prevention in Schools: specification, installation and
maintenance of intruder alarm systems*
HMSO (London): DES Building Bulletin 69

Department for Education (1993)
Crime Prevention in Schools: security lighting
HMSO (London): DFE Building Bulletin 78

Department of the Environment (1994)
Planning out Crime
DoE (London): Circular 5/94

East Anglian Church Security Group (1992)
Churches: a crime prevention strategy
East Anglian Church Security Group
A collaboration between Norfolk, Suffolk, Essex and Cambridgeshire
police crime prevention officers.

Ekblom, P (1986)
The Prevention of Shop Theft: an approach through crime analysis
Home Office (London): Crime Prevention Unit Paper 5

Ekblom, P (1987)
*Preventing Robberies at Sub-Post Offices: an evaluation of a security
initiative*
Home Office (London): Crime Prevention Unit Paper 19

Electricity Council (1988)
Essentials of Security Lighting: commercial and industrial premises
Electricity Council

England, N (1992)
'Security equipment'
In Williams, A (ed) (1992) *Specification 93*
EMAP Architecture (London)
An introductory discussion plus listings of security equipment and
suppliers. This publication is regularly updated.

Farringdon, D P, Bowen, S, Buckle, A, Burns-Howell, A,
Burrows, J and Speed, M (1993)
'An experiment on the prevention of shoplifting'
Crime Prevention Studies vol 1, pp 93-110

Fire Protection Association (1985-91)
Security Precautions
Loss Prevention Council (London)
Datasheets giving a good summary of current practice:

SEC1	Security equipment and systems (1990)
SEC2	Site layout (1990)
SEC3	Fences, gates and barriers (1990)
SEC4	External security lighting (1985)
SEC5	External CCTV systems (1990)
SEC6	Doors (1985)
SEC7	Windows and rooflights (1985)
SEC8	Access control (1991)
SEC10	Hardware for fire and escape doors (1991)

Fire Protection Association (1992)
Prevention and Control of Arson in Warehouse and Storage Buildings
Loss Prevention Council (London)

Football Stadia Advisory Design Council
Some FSADC guidelines are relevant to security. They are distributed by the Sports Council.

Griswold, D B (1992)
'Crime prevention and commercial burglary'
In Clarke, R V (ed) (1992) *Situational Crime Prevention: successful case studies*

Health and Safety Commission (1987)
Violence to Staff in the Health Services
HMSO (London)

Heywood, K (1994)
Light Pollution
Institute of Environmental Health Officers (London)
Security lighting is identified as one of the major sources of light pollution.

Home Office Crime Prevention Centre (1994)
Police Architectural Liaison Manual of Guidance (2nd edn)
Home Office Crime Prevention Centre (Stafford)
Prepared for use by police architectural liaison officers, this guide summarises CPTED (crime prevention through environmental design) concepts and provides some checklists for building types as well as some simple case studies. A useful cross-reference to the British Standards on security.

Hope, T (1982)
Burglary in Schools: the prospects for prevention
Home Office (London): Research and Planning Unit Paper 11
A significant study, relating crime statistics to building form and site layout.

Houghton, G (1992)
Car Theft in England and Wales: the Home Office Car Theft Index
Home Office (London): Crime Prevention Unit Paper 33

Hudson, M (1982)
Bicycle Planning: policy and practice
Architectural Press (London)

Humberside Police (no date)
Design Out Crime
Humberside Police (Hull)
Produced in collaboration with the Hull School of Architecture, this is a good attempt at communicating with architects. Summarising current ideas on security in design, it should perhaps be more generally available.

Hunter, R D and Jeffrey, C R (1992)
'Preventing convenience store robbery through environmental design'
In Clarke, R V (ed) (1992) *Situational Crime Prevention: successful case studies*

Jacobs, J (1961)
The Death and Life of Great American Cities
Random House (New York)

Jeffery, C R (1971)
Crime Prevention through Environmental Design
Sage (Beverley Hills, California)
The theoretical case for believing that the environment affects criminal behaviour, and that changing the environment will change criminal behaviour.

Johnston, V, Leitner, M, Shapland, J and Wiles, P (1990)
Crime and Other Problems on Industrial Estates
University of Sheffield, Faculty of Law

Jones, P H (1990)
Retail Loss Control
Butterworths (London)

Labs, K (1989)
'Deterrence by design'
Progressive Architecture November 1989, pp 100-103
A good, concise review of of situational crime prevention.

LaGrange, R L, Ferrado, K F and Supancis, M (1992)
'Perceived risk and fear of crime: social and physical incivilities'
Journal of Research in Crime and Deliquency vol 29, pp 311-334

Laminated Glass Information Centre (undated)
Glass & Architecture: Shopfront Security
LGIC (London)
Contact data: Laminated Glass Information Centre, 299 Oxford Street, London W1R 1LA tel 0171-499 1720 fax 0171-495 1106.

Laycock, G (1984)
Reducing Burglary: a study of chemists shops
Home Office (London): Crime Prevention Unit Paper 1

Lincolnshire, Nottinghamshire and Derbyshire Police (undated)
Garage Forecourt Security

Loss Prevention Council (1993)
Specification for Testing and Classifying the Burglary Resistance of Building Components
LPC (London): LPS 1175 Issue 2

Lyons, S (1992)
Lighting for Industry and Security: a handbook for providers and users of lighting
Butterworth-Heinemann (London)
A handbook for lighting, which covers many aspects of security lighting. However, it does not examine in a critical way the role of lighting in crime prevention, which may sometimes be overstated.

Marsh, P (1985)
Security in Buildings
Construction Press (Harlow)
A general guide written by an architect. Some attempt is made to integrate the findings from crime research.

Mayhew, P, Elliot, D and Dowds, L (1989)
The 1988 British Crime Survey
HMSO (London): Home Office Research Study 111

Mayhew, P, Mirrlees-Black, C and Aye Maung, N (1994)
'Trends in Crime: findings from the 1994 British Crime Survey'
Research Findings no 14 (Home Office, London)

McClintock, H (1992)
The Bicycle and City Traffic
Bellhaven (London)

MCM Research (1993)
Keeping the Peace: a guide to the prevention of alcohol-related disorder
The Portman Group (London)
Contact data: The Portman Group, 2d Wimpole Street, London W1M 7AA.

Museums and Galleries Commission
The Commission issues guidance notes on security. Contact data: Museums and Galleries Commission, 16 Queen Anne's Gate, London SW1H 9AA tel 0171-223 4200.

National Association of Health Authorities and Trusts (1992)
NHS Security Manual
NAHAT (Birmingham)
Contact data: National Association of Health Authorities and Trusts, Birmingham Reseach Park, Vincent Drive, Birmingham B15 2SQ.

Newman, O (1973)
Defensible Space: people and design in the violent city
Architectural Press (London)

NHS Estates (1994)
Design Against Crime – a strategic approach to hospital planning
HMSO (London): Health Facilities Note 5

Nichols, J E (1983)
A Guide to Hospital Security
Gower (Aldershot, Hants)

Painter, K (1988)
Lighting and Crime Prevention: the Edmonton Project
Middlesex Polytechnic, Centre for Criminology

Painter, K (1989)
Lighting and Crime Prevention for Community Safety: the Tower Hamlets Study (1st Report)
Middlesex Polytechnic, Centre for Criminology

Painter, K (1991)
An Evaluation of Public Lighting as a Crime Prevention Strategy with Special Focus on Women and Elderly People
University of Manchester, Faculty of Economic and Social Studies

Painter, K (1994)
'The impact of street lighting on crime, fear and pedestrian street use'
Security Journal vol 5, no 3 (July)

Petrol Retailers Association (1993)
Petrol Filling Station Physical Security Measures
PRA (London)
Contact data: Petrol Retailers Association, 201 Great Portland Street, London W1N 6AB.

Phillips, S and Cochrane, R (1988)
Crime and Nuisance in the Shopping Centre: a case study in crime prevention
Home Office (London): Crime Prevention Unit Paper 16

Poole, R (with Donovan, K) (1991)
Safer Shopping: the identification of opportunities for crime and disorder in covered shopping centres
West Midlands Police (Birmingham)

Poyner, B (1983)
Design against Crime
Butterworths (London)
An early review of research behind the emerging approach of crime prevention through environmental design.

Poyner, B (1992)
'Situational crime prevention in two parking facilities'
In Clarke, R V (ed) (1992) *Situational Crime Prevention: successful case studies*

Poyner, B and Warne, C (1988)
Preventing Violence to Staff
HMSO (London): Health and Safety Executive

Poyner, B and Webb, B (1991)
Crime Free Housing
Butterworth Architecture (Oxford)

Poyner, B and Woodall, R (1987)
Preventing Shoplifting: a study in Oxford Street
Police Foundation (London)

Ramsey, M (1991)
The Effect of Better Street Lighting on Crime and Fear: a review
Home Office (London): Crime Prevention Unit Paper 29
A demonstration of applying research scrutiny to the conventional wisdom about lighting and crime prevention.

Shop Front Security Campaign (1994)
Shop Front Security Report
British Retail Consortium (London)

Sinnott, R (1985)
Safety and Security in Building Design
Collins (London)
This book concentrates on housing, but some material is also relevant to non-residential buildings.

Sloan-Howitt, M and Kelling, G L (1992)
'Subway graffiti in New York City: "Gettin Up" vs "Meanin It and Cleanin It"'
In Clarke, RV (ed) (1992) *Situational Crime Prevention: succesful case studies*

Smith, L J F (1987)
Crime in Hospitals: diagnosis and prevention
Home Office (London): Crime Prevention Unit Paper 7

Sports Council (1990)
Designing for Safety in Sports Halls: Security
Sports Council (London): Technical Unit for Sport Datasheet 60.3

States of Jersey Crime Prevention Panel (undated)
Design against Crime
States of Jersey Police (St Helier, Jersey)
A well-produced compendium of current good practice in security design.

Stollard, P (ed) (1991)
Crime Prevention through Housing Design
Spon (London)

Strobl, W M (1978)
Crime Prevention through Physical Security
Marcel Dekker (New York)

Sussex Police (undated)
Security of Police Stations
Sussex Police (Lewes)

Sykes, J (ed) (1979)
Designing against Vandalism
Design Council (London)

Taylor Report (1991)
Hillsborough Stadium Disaster Final Report
HMSO (London)

Tien, J, O'Donnell, V F, Barnett, A and Mirchandani, P B (1979)
Street Lighting Projects: National Evaluation Program (Phase 1 Report)
National Institute of Law Enforcement and Criminal Justice (Washington, DC)

Tyne and Wear County Council Architects Department (undated)
Reducing Vandalism by Design and Management
Tyne and Wear County Council (Newcastle-upon-Tyne)

Underwood, G (1984)
The Security of Buildings
Architectural Press (London)
When published this was the only general guide to building security written by an architect. It still gives a good overview of traditional security concepts.

Vail Williams (1990)
The Occupiers' View: business space in the '90s
Vail Williams (Fareham, Hants)

Wallace, J and Whitehead, C (1989)
Graffiti Removal and Control
CIRIA (London): Special Publication 71

Ward, C (ed) (1973)
Vandalism
Architectural Press (London)

Webb, B, Brown, B and Bennett, K (1992)
Preventing Car Crime in Car Parks
Home Office (London): Crime Prevention Unit Paper 34
An easy-to-read paper showing how research can reveal the way that the crime risk in car parks is influenced by design and operation.

Webb, B and Laycock, G (1992a)
Reducing Crime on the London Underground: an evaluation of three pilot projects
Home Office (London): Crime Prevention Unit Paper 30

Webb, B and Laycock, G (1992b)
Tackling Car Crime: the nature and extent of the problem
Home Office (London): Crime Prevention Unit Paper 32

Wheeler, A H (1989)
Cycle Theft Update
Transport Research Laboratory (Crowthorne, Berks): Working Paper TS3

Williams, D (1989)
'Lessons in security' and 'Post-vandalism'
AJ Focus (Architects' Journal) June 1989, pp19–23 and 50–51

Wise, J A and Wise, B K (1985)
'The interior design of banks and the psychological deterrence of bank robberies'
In Pauls, J (ed) (1985) *Proceedings of the International Conference on Building Use and Safety Technology* (Los Angeles)
NIBS (Washington, DC)

Zeisel, J (1976)
Stopping School Property Damage: design and administrative guidelines to reduce school vandalism
American Association of School Administrators (Arlington, Virginia)

■